GROUNDED IN EIRE

GROUNDED IN EIRE

The Story of Two RAF Fliers Interned in Ireland during World War II

RALPH KEEFER

McGill-Queen's University Press • Montreal & Kingston • London • Ithaca

ISBN 0-7735-1142-3

Legal deposit fourth quarter 2001
Bibliothèque nationale du Québec

Printed in Canada on acid-free paper
Reprinted in 2001

McGill-Queen's University Press acknowledges the financial support of the Government of Canada through the Book Publishing Industry Development Program (BPIDP) for its activities. It also acknowledges the support of the Canada Council for the Arts for its publishing program.

National Library of Canada Cataloguing in Publication Data

Keefer, Ralph, 1952–
Grounded in Eire : the story of two Canadian RAF fliers interned in Ireland during World War II

ISBN 0-7735-1142-3

1. Keefer, R.G. (Ralph Gardener), 1917–. 2. World War, 1939–1945–Prisoners and prisons, Irish. 3. Prisoners of war–Ireland–Biography. 4. Prisoners of war–Canada–Biography. 5. World War, 1939–1945–Personal narratives, Canadian. I. Title.

D805.174K43 2001 940.54'72417'092 C2001-900263-7

This book was designed by David LeBlanc and typeset in 10.5/13 Sabon

To my father, R.G.C. (Bob) Keefer, DFC

It's not for the sake of a ribboned coat
  Or the selfish hope of a season's fame
That the Captain upon his shoulder smote –
  "Play up! play up! and play the game!"

Sir Henry Newbolt, *Vitai Lampada*
as quoted by Bobbie Keefer

CONTENTS

AUTHOR'S NOTE

As a Royal Air Force pilot during the Second World War, my father commanded over eighty missions. For nearly five years, from September of 1940 to April of 1945, he served in Bomber Command, Ferry Command, and Coastal Command in Europe and North America, and mastered just about every aircraft imaginable. Near the end of the war, he was even awarded a Distinquished Flying Cross for his efforts. Most would consider him a war hero – even if we didn't. To us he was just Dad, an unassuming guy who showed up every once in a while and paid the bills. Nonetheless, among the many thousands of men and women who served the RAF during the war, fewer than fifty survived as many missions as he did.

And you can't experience all of that, without picking up a few stories along the way.

One of my father's best – an unlikely story if there ever was one – involved a botched mission to Frankfurt where he and his friend Jack Calder, who had been a reporter before the war and was now his navigator, somehow landed up in an Irish prison camp. Not that any of us believed it, at least not at first. It wasn't exactly your typical POW yarn, not if your favourite heroes growing up, apart from him, were Steve McQueen in *The Great Escape* and Colonel Hogan in *Hogan's Heroes*. In fact it would have remained a family secret for all time had it not been for his decision to turn it into a book. That process began in 1985 and finished, we thought, in 1990, on his seventy-fourth birthday, with the completion of a manuscript tentatively entitled *My Secret Irish Escapade*. Little did we know. For by then the earliest signs of Alzheimer's had set in, and the story had obviously suffered in the telling.

Not like in the early years, when the empty bottles would gather between us and it seemed like *maybe* he was telling the truth.

That's where I came in. With three children of my own and a full-time job, I didn't have a lot of time to devote to the project. I did review the original manuscript; however there was also an old shoebox full of photographs, newspaper clippings, and reports that helped immeasurably. Still, it seemed an impossible task, with so many unanswered questions. Feeling guilty, I then took a leave of absence from my job and went to Ireland. Fortunately, I met a lot of great people there, including a few of his old friends and captors, each of whom added additional bits to the story as only the Irish can. I then rewrote the whole of it, from start to finish. It's a wonderful story, really, unique in the annals of history, as he used to say. And with Alzheimer's disease being what it is, it soon got to the point where working on the book was the best way to remember him.

So, in the end, this book is very much a father and son effort. It's a combination of his story and my words, imperfect as they may be. I don't know how else to explain it, Dad, other than that.

On the other hand, this book is the result of more than *just* misplaced pictures, forgotten memories, or idle family reminiscences. Thanks are due, in no special order, to *Maclean's* magazine and Canadian Press for permission to reprint Jack Calder's articles; to Bruce Girdlestone for permission to use his 1943 paper *Prisoner of the Green*, a copy of which he provided my father in the early 1980s; to Aubrey Covington, Bud Wolfe, Chuck Brady, and Mr Girdlestone, once again, for their friendship and many funny stories over the years; in Ireland, to Commander Peter Young of the Irish Military Archives in Dublin, Commander Bill Gibson of the Irish Army stationed at the Curragh, Martin Gleeson of the Irish Warplane Research Group, and Tomsy O'Sullivan and Cyrus Vance in county Clare. I particularly enjoyed the pub crawl, Tomsy – I know he would have too.

Closer to home, my everlasting gratitude to Mike McCormack for his many helpful comments on Irish history and the war; Mark Rash, my travelling companion and erstwhile golfing partner; Walter

"Punch" Thompson, DFC, a family friend who helped immeasurably in understanding RAF terminology and culture; Fionuella O'Byrne Terri Rear, and Louise Keefer for their work on the manuscript; John Esson for his out-of-print book on de Valera which I'll return some day, I promise; Terry Harris for his computer expertise; Steve Barley, whose airplane I nearly crashed trying to understand all of this; and Steve Harrison and Peter Cullen, two Irish expatriates and friends who, like my father, love to tell stories.

And last and certainly not least, thanks are due to Philip Cercone, the director at MQUP, who stuck by the story all these years; to Joan McGilvray, the coordinating editor there, a wonderfully patient woman; and to the others who assisted in the production of this book.

Your efforts were greatly appreciated, and we couldn't have done it without you.

GROUNDED IN EIRE

1

THE TRIP OVER

Lost, low on fuel, and fearing the worst, Pilot Officer Bobby Keefer ordered his crew to assemble by the hatch. Jack and the other three stepped forward, but not Virtue. Damn it, thought Keefer, not Virtue.

Not again.

Keefer called his tail gunner on the intercom one last time. No response. Checking his watch, he turned the controls over to Diaper, his second pilot, grabbed the axe at Jack's feet, and headed back. Arriving at the entrance to the rear turret, he looked down. The outer of the two turret gates was jammed open, sticking out of the aircraft, while the inner was jammed shut, blocking his path. As the cold air swirled at his feet, he looked inside. There sat Sergeant Virtue, their tail gunner, alive, but frozen in his perch. Poor sod, no wonder he hadn't come forward. It had to be -30° C in there!

In a panic, Keefer quickly took the axe to the closed gate only the hinges proved stronger than expected. He cursed again; the timing couldn't have been worse. After a dozen or more blows, he managed to split the lower of the two hinges, ripping the gate off its post. "Hit the silk kid," he shouted, grabbing a stunned Virtue by his collar and dragging him through the opening. Now at least the kid would have a chance.

Keefer rushed back to his seat. He stared out from the cockpit, straining to see if the view had changed any since his return. It hadn't. Darkness remained, and the moving shades of gray and black continued to cross their nose, too dense to suggest an opening above or below. He listened to the rain, thundering in sheets against the aluminum and cloth skin of their small, twin-engine Wellington. He prayed.

He looked at his watch again. It was 05:15 – ten minutes past their estimated time of arrival. Had he been gone that long?

If they survived Keefer knew he would have some explaining to do – not only why he hadn't noticed that the gate was damaged earlier, but why he had gone back himself, leaving the controls to his second dickie, a sprog like Virtue on only his second mission. They were a team after all, and his responsibility was at the helm, not scrambling around in the back playing the hero. But surely the Old Man would understand, given what had happened to his previous tail gunners. And if it meant trouble with the brass, then so be it. At least they would be home in one piece.

Keefer's eyes returned to his watch, only this time his tired hands were shaking, and he could barely see it on his wrist. It had been that kind of a mission.

The mission had begun, as most World War II nighttime bombing missions began, on the evening of the previous day, in this case 24 October 1941. Donning the same scarf and shirt he always wore (bomber pilots were a superstitious lot) Keefer had woken early, just in time for a late breakfast. He met Jack in the mess, where they were introduced to their newest crewmember, Sgt Virtue.

Their last rear gunner, Sgt Cox, had baled out two days earlier over Holland. That was Mannheim. Tonight was Frankfurt.

Following a vague and ultimately unreliable weather brief – 5/10 cloud decreasing to stratus over the Dutch coast with a ceiling to 7,000, and clear over the target – they had assembled on the tarmac. Queued before the Hampdens and after the Whitleys, it was to be a "maximum effort" that night, a joint operation among three squadrons, including his, the 103, stationed at Elsham Wolds in Lincolnshire. This was part of Bomber Command's response to the Battle of Britain the previous year, hammering medium-range German targets en masse despite huge losses (one of every four crews) until 1943 when the larger, four-engine aircraft took over. They didn't get off as planned, however. Virtue had taken his guns out for cleaning following a test that afternoon, only he'd lost a

Clockwise from top left: Sgt Dalton, The Old Man, aka Wing Commander Freddy Ball, Sgt Tett, Pilot Officer Keefer, Sgt Brown, Sgt Cox

Sgt Virtue

Sgt Diaper

The 103 crest with its motto, "Touch me not."

part. That was all well and good, thought Keefer, until the bin rat had failed to show up.

"Christ, Virtue. Who gave you permission to do that?"

"You did, sir. Sorry about that. Wanted to be at my best, sir."

"Jesus, Virtue, warn me next time." If there was a next time. That's what got him about the English. They were all so bloody nonchalent.

It must have been in the rush after that, he realized, that Virtue had forgotten to seal his gate.

And then, sure enough, the clouds had thickened over the North Sea, just as he had expected, just as they all expected, for they had been flying in them most of that month. So instead of being part of a deafening wave of death thundering out over the North Sea, he would now be on his own, straggling at the back. In the clag. In no man's land, unable to turn back (no one did that) or continue, at least not without a good chance of biting it. As he approached Mannheim, the Wellington had then iced up, just as it had two nights earlier when they had lost Coxie. Coxie, his last tail gunner, who hadn't answered the intercom either.

And then, to make matters worse, he had waited until 02:30 before finally issuing the bombing command. According to his log, he had chosen an unidentified town along the Rhine. This, he figured, was half way, although in reality he had no way of knowing except by fuel consumption. It was often like that during the war, he had learned, bombing God knows what at the last moment in order to stretch his remaining fuel back to base. It wasn't his fault that the Wellington hadn't any range to it.

Just after Jack completed the drop, they were hit by flak – pencil ribbons of purple, yellow, and orange streaking up from the ground, exploding at their altitude. One burst rocked the front of the fuselage, disabling their loop antenna, while another nailed them broadside where Jack normally sat behind him at the navigator's desk, sending his friend flying from his seat, his curtain wrapped around his neck as if he was trying to hang himself. If he had known where Jack was taking them, Keefer later joked, he would have saved him the trouble.

Keefer, bolt upright in his chair, immediately tightened his grip on the stick. It was never a good thing to be last to the party, not when the hosts were armed and dangerous. He began jinking, an evasive maneuver designed to avoid a steady, predictable course. But the tracers kept coming, as the Wellington bobbed and weaved its way through the sky. He worked the throttles to desynchronize the motors, hoping to confuse the groundfire, counting the seventeen seconds that it took for an 88 mm ground shell to reach a bomber at 20,000 feet, and then holding his breath. Like waiting for the lightning, after the thunder had sounded.

Still, he managed to jink his way out of it until they escaped the action, and fixed a 278° reading for home. A hollow victory, since with no antenna they would have to fly home blind or by dead reckoning (DR), as it was through the antenna that they received beacon signals from the English coast. He had flown DR from Germany twice before on shorter missions to Essen and Munster, though never from Frankfurt. And never in this weather.

So, as the hours passed he had plenty of time for reflection, flying alone in the dark, knowing he was lost. Assuming he'd be next. What would they say about him, he wondered. A fine pilot who, like his father, a member of Canada's Imperial Air Force who had perished in a training accident in Virginia in 1917, had died before his time. A former football and hockey star who had won his two inches worth in the Montreal *Gazette*, Toronto *Star*, or the Ottawa *Citizen*, and then had been quickly forgotten. Like Coxie, his last tail gunner. Or Brandon McCabe, the one before Coxie, who had lasted two missions before a Messerschmitt nailed him in the stomach. Or Jimmy Nesbitt, his first, who had died over Emden. Jimmy used to throw up on the runway – like a goalie on the eve of the Stanley Cup final. Tail gunners, crazy, each and every one, spending the war ass-backwards in a two-cubic-foot-glass icebox, suspended in mid-air like a sitting duck, like a carnie waiting for some brat to get lucky with a throw. Anyone in their right mind would prefer to be a nose gunner, he thought, like Sgt Brown, at twenty-nine, the old man of their crew. Brownie may have been a gunner, but at least he was pointed in the right direction.

Anyway, that's what made him pick up the ax. So damn the Old Man.

Keefer steadied himself and pulled back on the stick. The altimeter swung back – 3,000, 2,000, 1,000 – and still no break in the clouds. He cursed again, he should have done this twenty minutes ago.

Keefer reached for his headset, cupping the microphone in the palm of his hand. "Mayday ... mayday ... mayday ... C for Charlie C2506 ... Over." He knew that he was breaking the rules – emergency calls were prohibited for fear of disclosing transmitter locations – but sometimes rules were made to be broken. He switched to receive. No response.

Jack remained seated at the navigator's desk, examining the Gee charts, air almanac, navigation books, and calibration documents strewn before him. He seemed puzzled, reacting slowly to the situation. He got up, shuffled over to the pilot's seat, and lifted the earflap on Keefer's leather helmet. "Ya know, Bobby, we might still be over England," he shouted. "Frankfurt to base, four-five-zero; air speed, one-four-zero. Tail wind maybe two-five, zero-twenty." Jack then paused in mid sentence and seemed to change his mind. "But then, with a zero fifty, we might be over Wales."

Keefer looked up at his friend. "Thanks, Jack." Or more likely the North Sea, he mumbled to himself, for he had faced strong westerlies on this route before. Too late to worry about that now. He repeated his first S.O.S. Still no response. Then he thought for a moment. Maybe the winds were off. Maybe Jack was right. "Do we have a map of Wales?" he shouted, as his friend turned back towards his desk.

"Yeah," Jack replied over the intercom a moment later, "but it ends at Aber ... Aberysth ... Abershwetze ... Aberystwytheee ...

"How far past Elsham is that?"

"I'm not sure," Jack replied.

"Christ, Jack, you're the navigator."

Jack was silent. For the first time he appeared worried. After bombing missions to Boulogne, Karlsruhe, Munter, Cologne, Duisberg, and a dozen other German towns, Jack knew Germany pretty well – he claimed to recognize every pair of tits south of Frankfurt – but getting to that point had taken a while. Like Keefer, Jack had some catching up to do. Not that Pilot Officer[1] Jack Calder, from Goderich, Ontario, was dangerous or anything; Keefer would never have said that about his best friend. Or that placing volunteer officers seconded from the colonial air forces in charge of English Noncommissioned Officers (NCOs) was an ill-advised policy since they were, after all, university graduates. (McGill in Keefer's case, University of Toronto in Calder's). It was just that neither had spent much time *west* of their base.

"Mr Keefer, sir. Mr Keefer, sir!"

Keefer turned in the direction of Sgt Dalton, his wireless operator, who was yelling at him from the hatch.

"Sir I couldn't help overhearing you and Mr Calder, sir ..."

Sargeant Dalton, whom they called Dolly, had overheard part of the conversation between his two superior officers and was obviously concerned.

"Yes, Dolly," yelled Keefer over his shoulder.

1 Newly commissioned officers in the RAF were either pilot officers or flying officers, regardless of whether they actually went on to become to pilots.

Keefer's first solo in a Tiger Moth, No.6 Elementary Flying School, Prince Albert, Saskatchewan, February 1941. Sgt Dalton trained south of there in Regina.

Graduation day, 17 March 1941, as luck would have it.

"If I can be of assistance, sir." The young sergeant stepped closer to Keefer's ear. "West of Swansea is the Irish Sea which stretches one hundred miles to the west. Then there's Ireland. If we were riding a strong easterly, sir, we may be that far." Although Dalton, like Virtue, thought of the two Canadians as reasonably approachable compared to some of the English officers he had met, their higher rank nonetheless endowed them with a wisdom that the young sergeant sometimes found intimidating. Otherwise he would have said something earlier.

"How big is Ireland, Dolly?"

"You don't know much about this part of the world, do you, sir?"

"No I don't, Dolly. And I bet you don't know where Saskatchewan is either."

"It's between Alberta and Manitoba, sir."

"Dolly, answer the fucking question."

"Yes sir." The young Englishman seemed to be making progress in conquering his feelings of inadequacy. "I'd say Ireland is one hundred and fifty miles at its widest. From north to south it's a little more complicated." Keefer knew that much. "Yes, sir, just in case we overshot, sir. The South broke off, sir, as you know. It used to be called the Irish Free State but they changed that three years ago. Now they call it Eire, sir. Yes sir, as in hair. And de Valera, their leader. He's a right bastard, he is, sir. U-boats are attacking our ships off their coast and many of our navy lads have been lost at sea. We keep expecting the Irish to join us but so far no luck. If we really are over the South, sir, they probably won't answer a mayday. In fact, sir, I've heard tell they have a camp. Thought you better know that, sir."

Keefer took a moment to compose himself. Ireland for chrissake! That can't be right, we can't be that far! Still, stranger things had happened, he knew that much. Guys overshooting or coming up short, landing in fiords, or bogs, or fields if they were lucky. Flying DR in this weather was anyone's guess.[2]

It was now 6:06 and the crew was silent and tense. While Keefer adjusted his Mae West, they each adjusted theirs, scouring the darkness one final time. They were all preparing for the worst.

It was then that they caught a break. At the last moment, Sgt Diaper, the second pilot, spotted a faint coastline parallel to their course, three miles to the west. The Wellington's engines began to sputter - first the starboard, then the port.

With a deep sigh of relief – and now knowing they weren't going to die after all – Keefer cut the master switches. He asked Sgt Dalton to release a Verey signal cartridge to warn those below, and then ordered the crew to bail out.

A moment later the four Brits were crouched at the open door. Like Keefer, no one had actually jumped before, so every one was a little nervous. Everyone but Jack, who was ushering them out with a typical flair for the dramatic. "Cheer up lads," he announced. "Last one down buys the first round. If its England, its Newcastle; if its Ireland, it's Guinness." The four sergeants rolled out in turn, disappearing quickly into the darkness.

Finally, Keefer began to relax, relief draining from his limbs. No matter where they were, he thought, things could be worse. Then he saw the lights. Even a few seemed noteworthy coming from the blackouts in England. So Ireland it is!

2 Bomber command diaries for 24/25 October 1941, for example, indicate that of the seventy Wellingtons, Whitleys, and Hampdens dispatched to Frankfurt, only eight reported reaching the target. Four failed to return, better than average as it turned out.

Year 1941 Month	Date	Aircraft Type	Aircraft No.	Pilot, or 1st Pilot	2nd Pilot, Pupil or Passenger	Duty (Including Results and Remarks)
—	—	—	—	—	—	— Totals Brought Forward
Oct.	22	Wellington	2506	Self	Sgt. Diaper P/o Calder Sgt. Dalton Sgt. Brown Sgt. Cox	Operations To Mannheim (13) Bombs - 1x1000 - 4x500 - Nickels. Iced Up at 18.200 ft. inside Dutch Coast - Recovered at 10.000 ft. Went down to 3000 to get below freezing level. Bombed Dutch Aerodrome at 2000 ft. Cox baled out. Large hole in bottom of aircraft. Electrical storms on way home. Assumed landing place of Sgt. Cox is near Antwerp.
Oct.	24	Wellington	2506	Self	Sgt. Diaper P/o Calder Sgt. D. B. + Virtue	Air Test
Oct.	24	Wellington	2506	Self	Sgt. Diaper P/o Calder Sgt. Dalton Sgt. Brown Sgt. Virtue	Operations To Frankfurt (14) 10/10 Cloud. Bombed Unidentified Town on Rhine. Entire Trip D.R. Reciprocal Radio bearings. & eventual complete radio failure. Navigator later reported concussion. Severe Icing Clouds. Identified position over Eire with 10 mins petrol. Headed north Engines Cut - Bailed Out - All of Crew Interned in Eire.
					Grand Total [Cols. (1) to (10)] 332 Hrs. 55 Mins.	Totals Carried Forward

Last page of Keefer's log prior to his internment. The "complete radio failure" noted in the 24 October entry should have been revised, as the reader will discover in the next chapter.

By then his altimeter had fallen to 720 and the rain had dried to a scotch drizzle. There were broken clouds to the west and light filtering through, with what looked like a farmhouse at the top of a hill. It looked so easy and treeless that jumping no longer concerned him. He even considered a forced landing, although with no power he knew they wouldn't stand a chance.

He then looked over at his friend one last time. Maybe they would make it through the war after all, he thought, fastening his helmet with a snap and betting that they'd be back in Lincolnshire by nightfall. Maybe this friendship would last, unlike so many others. His mother certainly hoped so, judging from the letter that

Keefer and Calder at Corries, near Churt, in Surrey, England, August 1941.

Pilot Officer Jack Calder

he Gazette

MONTREAL, MONDAY, APRIL 28, 1941.

CANADIAN REINFORCEMENTS FOR SKY WAR

Six Montreal airmen, whose safe arrival in Britain was announced yesterday, prepare to board the transport in w… they were convoyed overseas. From left to right: PO. Ralph G. (Bob) Keefer, PO. W. L. S. (Bill) O'Brien, PO. Edm… T. Asselin, PO. H. E. Reilley, PO. Herbert E. J. Whiston, PO. Leslie R. Farrow. The first three are pilots and the ot… observers.

Gazette Photo (Copyright Reserved

LARGE CONTINGENT OF AIRMEN ARRIVE

(Continued from Page One)

True to the tradition that once a newspaperman always a newspaperman, Pilot-Officer Jack Calder of Goderich, Ont., former member of The Canadian Press staff at Toronto, appeared on scene with notes jotted down during the trip.

"Always like to help out an old friend," he said, adding with a smile: "And all I want to know is whether you can tell me if the sunset here is as beautiful as sunset over Lake Huron."

Above left and right *Montreal Gazette*, 28 April 1941

King, Queen At Windsor Repay Canada Hospitality

Entertain Fliers From Dominion—"Think You Can Trust Me With Military Secret," Her Majesty Tells Goderich Pilot Officer

Windsor, England, April 28—(CP)—Newly-arrived Canadian airmen were surprised while walking in the grounds of Windsor castle by the sudden appearance of the King and Queen and Princesses Elizabeth and Margaret Rose.

A short time earlier their majesties had been elsewhere nearby, but when they were informed of the Canadians' visit they motored with the princesses promptly to the castle. There they greeted their visitors and expressed the desire to reciprocate "for the kindness shown by everyone" during their tour of Canada in 1939.

While comrades were introduced to their majesties by Pilot Officers Jack Calder of Goderich and Bill O'Brien of Montreal, the princesses made their own acquaintances among the group, composed of pilots, observers and gunners. They gossiped generally with one group. Pilot Officer Hugh Miller, Windsor, N.S., asked the girls if they preferred the red coats of the Royal Canadian Mounted Police or the R.C.A.F. blue uniform.

"Dentist Worse Than Raid"

Princess Elizabeth replied discreetly as if unable to make up her mind, giving a non-committal: "Oh, I don't know." But Margaret Rose came right out with "I like the red-coats best. They're brighter, aren't they."

The girls expressed regret they were unable to visit London, explaining: "They say there's nothing there for us now except the dentist."

"And he's about as bad as an air raid," sympathized Calder.

"I think he's worse," declared Margaret Rose.

While the Canadians were walking through the grounds, Pilot-Officer Clifford Chappel of Windsor, Ont., who was beside the King, commented upon the beauty of the flowerbeds. The King agreed, but said next year the flowers would be displaced by vegetables.

"You Can Trust Me"

After tea his majesty apologized for having to return to duty, but the Queen remained a while longer and asked Calder, as senior officer of the group, how many men came to Britain in his crossing. Calder was about to reply when on second thought he apologized, saying, "I'm sorry, but guess I shouldn't say seeing it's a military secret."

The Queen appreciated his caution but lent assurance by saying "Oh, I think you can' trust me." And he did.

Her majesty recalled incidents of the royal tour and told her young guests:

"It's so nice to be able to think of something so pleasant as the times we had in Canada, when we are going through such troublous times now. Everyone in your country was so good to us."

Toronto Star, 28 April 1941

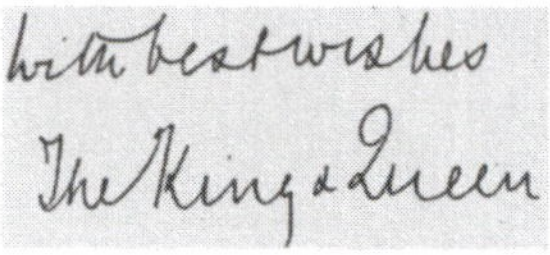

Royal salutations.

Keefer carried in his left jacket pocket. She was delighted to hear that the two were doing so well, she wrote, impressed that they found time to relax between missions. She had also been pleased to learn that they had visited the cousins in Surrey and had actually taken the girls out, a redhead for Jack and a brunette for Bobby. She had been surprised, to say the least, that they had then driven to Scotland in Uncle Wilks' '35 Hudson convertible which Keefer had bought for £10 and sold a month later for £15 when the insurance came due. The boys had apparently managed to coax some petrol tickets out of the Old Man with a large bottle of Uncle Wilks' Canadian Club rye – no mean feat in the middle of a war! And she had remembered Jack from the previous year, of course, when the two had come to Ottawa for thanksgiving dinner. The whole thing made her proud, she concluded, even if Jack's friends at Canadian Press hadn't seen fit to mention him! Proud, not just of how famous they had become back home – picking up the newspapers and seeing their photographs there, of how they had actually met the king and queen at a reception for the many volunteers arriving by boatload that year – but that they had volunteered at all, that they had answered the call. Yet worried, as any mother would be, that she might never see him again.

No matter what followed, thought Keefer, they'd find a way.

"Do you think Dolly is right?" he asked as they stood by the door, feeling the chill wind on their faces. "Do you think we'll be detained?"

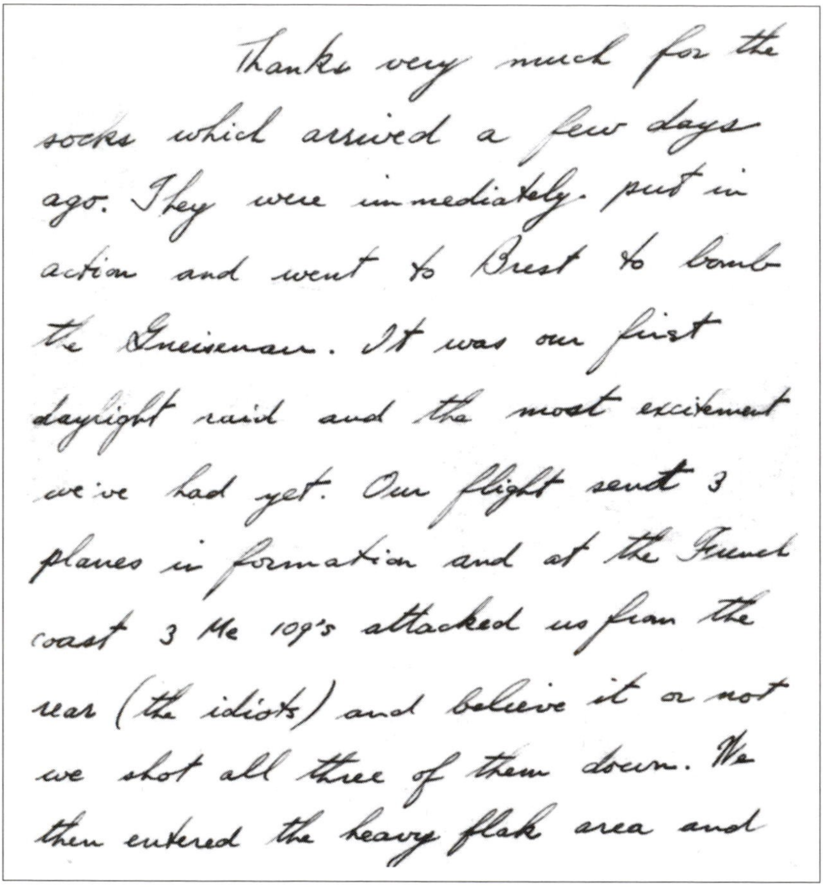
Thanks very much for the
socks which arrived a few days
ago. They were immediately put in
action and went to Brest to bomb
the Gneisenau. It was our first
daylight raid and the most excitement
we've had yet. Our flight sent 3
planes in formation and at the French
coast 3 Me 109's attacked us from the
rear (the idiots) and believe it or not
we shot all three of them down. We
then entered the heavy flak area and

An excerpt from one of Keefer's letters home.

"Are you losing your senses, too?" Jack replied, checking his chute. "The Irish are the friendliest people in the world." Jack was thinking of those he had met back home in Ontario, in Goderich or dozens of other Ontario towns like it. The Irish who had arrived by the boatload up the St Lawrence in the middle of the last century, who had built the railroads, mined the ores, and grown the potatoes, who had served Canada so well. Or those before them, the Empire loyalists who had fled the Republicans to the south, journeying north across the Niagara peninsula to the Great Lakes where Jack, the son of an Empire loyalist and Anglican minister, had spent summers with his family.

Not necessarily of those who had remained.

"Even if they're neutral, they'll just take us to the border and release us."

"Then, why do they have an internment camp?"

There was a brief silence as the two Canadians thought about that one.

"Well, even if they do lock us up, I can't see it lasting more than a few days. You know, straighten out the paper work, that sort of thing. What are they going to do? Lock us up in a bloody brewery?" Jack's eyes were bright, with a smile creasing the corners of his mouth. "But if they did, what a story that would be."

"Always the newspaper man, eh Jack." To Keefer, Jack would always be part reporter, even if he claimed otherwise, even if he claimed to have given it up in the name of a good cause. No doubt about it, thought Keefer, remembering Jack's version of their raid on the *Gneisenau* in August, the guy could shoot a line with the best of them. Jack had joined their New Zealand friend, Slapsy Maxie, on one of their few daylight missions, ending up on the front pages of many of the newspapers across Canada and even a few in the US as well.

Jack smiled again. When he did, his fair-skinned complexion seemed to redden, highlighting his freckles, turning his auburn hair closer to red, making him look younger than he really was. In fact, Keefer often thought that Jack looked Irish, or at least like Keefer's

'I BOMBED THE GNEISENAU' DEATH CLOSE EVERY SECOND

Slapsy Maxie Cool So Everything Came Off All Right

STARTED WITH LEAK

By
PILOT OFFICER JACK CALDER
of Goderich
(Copyright)

London, Aug. 13—(CP)—Slapsy Maxie looked at the oil pressure again and then he looked at me.

"I think we might just make it," he said.

Then it was time to dash out to sea. We climbed three minutes through broken cloud. Behind were other formations—all streaking for Brest and the Gneisenau.

We suddenly broke cloud. Below was the sea, blue, and far ahead was France.

I thought of that bad oil leak which might make our gun turrets unserviceable at the target area. The wireless operator was working to repair it. His wedding was only a couple of weeks ahead and he wanted to come back for it.

Objects on the coast grew bigger and bigger and I went forward to the bomb-aimer's position. Max patted me on the shoulder. He had refused to turn back because we were leading the formation.

Ack-Ack Comes Up

"Look at that ack-ack come up," some said.

Ahead of us the sky was cluttered with little patches of anti-aircraft shell bursts. We headed across the little peninsula toward the drydocks of Brest.

"I'm going to try the run-up from here," Max 'phoned.

"But we're 10 miles off the target and will be there two minutes early," I told him.

"That's all right," he argued. And he was captain. "We've got a good chance of getting in."

Below I saw barrage balloons at about 6,000 feet. Something was burning on the ground—probably a fighter aircraft.

It happened like a flash.

"Rat-a-tat-tat-tat-tat," went the guns in our rear turret.

"I just shot down an Me109," called the rear gunner. "I didn't even have time to report it."

"Heads up," cried the gunner leader. "Here come two more."

"I've got one," called the rear gunner of No. 3 aircraft.

"Let's have less people talking at one time," Maxie ordered. "All right, but I've got an ME109 too," yelled No. 2 rear gunner.

Victim Bails Out

Below us I saw a parachute open. One victim had managed to bail out. Someone reported one fighter had crashed.

Now we were doing violent evasive action. The ack-ack was coming at us.

"Broomph! Broomph!" went a burst right beneath my face. I put on my parachute.

"You'll have to do a 90-degree turn left to make it," I told Maxie. "Quickly."

He wheeled her over and the formation followed.

"Left, left," I called. "Steady."

The target was coming down the drift wires.

"Dropping, dropping, go," I hollered and the bombs sailed out. On the last turn No. 3 had slithered above us and his bombs nearly hit us as they fell.

"Bombs gone," I shouted.

"Good show," said Maxie, "Now let's get out of here."

At that moment I knew we had missed the Gneisenau because I had felt a slight skid as we came out of the last sharp turn.

As we dived and turned, I saw the bombs bursting slightly beyond the target. At least we had helped to mess up the docks.

The shells were breaking all around us. We got out because Slapsy Maxie was so cool. At the

JACK CALDER

coast Maxie put the nose down and streaked for home.

Wellington Struggles Home

He throttled back when he saw a damaged Wellington struggling home on one motor. We stuck with her until our fighters came along to protect her halfway across the channel.

We landed in the south of England to refuel. I crawled out, had tea in the officers' mess and lay down on the grass.

But when I closed my eyes all I could see was a map of Brest, the drydocks and the Gneisenau. A lot of people got her that day.

The next day a friend from Montreal and I were playing golf when we met Maxie and his wife.

"Did you have much trouble yesterday?" Maxie's wife asked, looking at me a little anxiously.

"Oh, we had some fun——," I started.

"Don't believe him," protested Maxie. "He's an awful line-shooter. It was very quiet."

He stared at me quite hard, as if to say: "I wouldn't tell her. Don't you."

Our formation of three has been credited officially with three enemy aircraft.

Toronto Star, 13 August 1941. Versions of this article appeared in most Canadian dailies, as noted.

stereotype of what an Irishman looked like, particularly after a few pints, with his red hair, red face and upbeat manner.

"Either way, Bobby," said Jack, who was now studying the coastline, "we're better off ditching in case we need to make a run for it." Ditching, in this case, meant setting the trim tabs for the coast, so that the Wellington would avoid crashing on land.

"Where are we going to run to?" Keefer asked doubtfully, fixing his tabs due west.

"The north."

"You mean, Belfast?"

"Yeah, Belfast. Hell, Bobby, with your broken-field running style, it'll be a piece of cake."

Jack, for his part, had been impressed when he had met Keefer three years earlier, on a cold, windy afternoon in Montreal. As the sports editor for Canadian Press, he had come down from Toronto on the train to cover the big game, one in which Keefer's team, the McGill Redmen, had defeated the University of Western Mustangs 9-0 to capture their first national football championship in over a decade. Back when the college games outdrew the professional ones. Crazy-legs Keefer, they called him, as the weakside back in McGill's victorious single wing formation that year. Jack had moved on to the editor's desk the following spring, but it certainly had been a night to remember. Especially the big riot afterwards when a few hundred McGill students, in their enthusiasm, had stormed the German Harmonia Club on Drummond Street after Hitler had supposedly "fined" the Jewish community in Berlin $400 million earlier in the week, the same week as the *Kristalnacht* riots in Austria, where Jewish shop windows were smashed, synagogues looted, and thousands carted off to concentration camps. A lighter moment in history, to be sure, overshadowed by the darkening clouds of holocaust, and war.

Calder, on leave with Keefer in London, September 1941.

McGILL GAME RIOT TO BE PROBED TODAY

University and Students Council to Investigate Victory Demonstration

THREE ARE ARRESTED

Arts Men Charged After Mob Invades German Club, Does $300 Damage—Several Scuffles Occur

Two investigations—one ... Gill University officials a... other by the executive co... of the Students' Society—wi... ably be undertaken today a... sult of the property damage ... Saturday evening followin... success of the McGill footbal... in winning the Canadian inter... giate championship, its first ... years.

The centre of the "riot" wa... German Harmonia Club at ... Drummond street, where, ... claimed, the damage amounte... $300.

Once in scoring pos... the Reds made no mistake. Hamilton zipped a forward pa... Bob Keefer that went for a ga... 15 yards and put McGill in po... sion on the Varsity seven-yard ... Bob Kenny plunged for four y... and on second down Anton smash-ed over for a touchdown. Keefer converted neatly from placement and the score was 6-1.

Again a fumble put Varsity in a scoring spot and the Blues second point followed. Once more the ... gan a downfield march, ... by a 40-yard dash by ... broke clear on an ex-...

BOB KEEFER
Half back

Gaining almost at will on the ground, the Redmen rolled seven touchdowns over the R.M.C. line with big Bob Keefer, outstanding on the backfield, scoring tw...

* * *

After the game, Coach Kerr thought Keefer's departure from the game was fishy. Keefer and Massey Beveridge are staunch pals and Beveridge took Keefer's place. The way Roberto was ripping off the tape after the game with no twinges of pain, his alleged late-game injury seemed a little like brotherly love to get Beveridge into the title fracas. In any event there ...

...Gill Champion Grid Team
...ils Coach Kerr After Game

THE GAZETTE, MONTREAL, MONDAY, OCTOBE...

McGill Beats Queen's, 17-5, for 3rd Straight

LOCAL TEAM HOLDS 1ST PLACE IN LOOP; WESTMAN SHINES

Retains Sole Possession of Leadership With Western 2 Points Behind

BOB KEEFER CASUALTY

Red Halfback Suffers Shoulder Injury That May Keep Him Out of Action for Some Time

MCGILL'S REDMEN TROUNCE TRICOLOR BY 17-5 BEFORE 12,000

VOL. CLXVII. No. 26...

THE GAZETTE, MONTREAL, MONDAY, NOV...

McGill Trounces Varsity, 23-2, as 14,800 Fan...

...ELIMINATES BLUES FROM TITLE QUEST; ...D EXTEND STREAK

...man Score Fifth Straight ...ictory in Intercollegiate Football

...LY-4 TOUCHDOWNS

ANDY ANTON SCORES TOUCHDOWN AS REDS SINK VARSITY HOPES

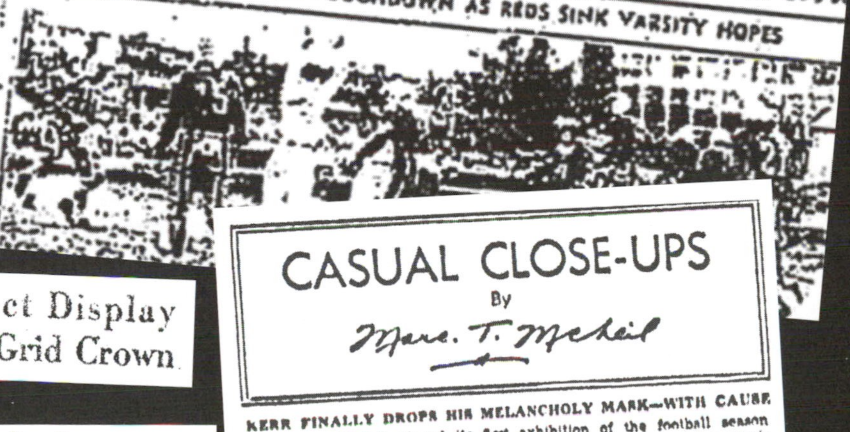

Near-Perfect Display Gives Reds Grid Crown

* * *

Bob Keefer's shoulder stood up for all but one minute of the game. Robert looked quite his usual self on the McGill back division.

* * *

CASUAL CLOSE-UPS

By

Marc. T. McNeil

KERR FINALLY DROPS HIS MELANCHOLY MASK—WITH CAUSE

Before McGill played its first exhibition of the football season against the Cubs, Doug Kerr, the Reds' head coach, sounded a pessimistic note. Said he in an ominously low voice and with a morose shake of his head, "Bob Keefer won't be able to play against Cubs, and that means our timing will be all off. Everyone takes their timing from Keefer, and so I don't know what will happen." (P.S.: As you recall, McGill won that game in a canter, 14-3).

Saturday afternoon, just after the final whistle sounded, Kerr emerged from the coaching box high above Molson Stadium's gridiron. He looked more woebegone than usual. Spying us, he said, "What did I tell you? See what happened after Keefer was hurt today, our timing went to pieces," and he dolefully headed for the ladder leading down to the ground. To look at him you'd have thought McGill had just absorbed a 30-0 thrashing instead of having achieved a decisive and glorious triumph over Queen's, 17-5. It was on the tip of the tongue to make some remark about having often heard of the eternal optimist but of never, until that moment, having encountered the eternal pessimist, when Kerr turned back suddenly and said, "How did you like that fake kick play?" Then he departed quickly, wearing the first grin he has succumbed to this season.

Montreal Gazette, 24 October, 7, 14, and 21 November 1938

Not to mention Keefer's first hint that odd things might happen whenever he and Jack got together.

"OK, where's Belfast?" Keefer asked. There was a Royal Air Force base near Aldergrove, almost directly due west of Elsham, though neither had ever seen it.

"I'm not sure but it can't be that far. Ireland's not a very big place, you know."

Keefer scoured the ground below. It didn't look very big to him either, at least not at that point. If they overshot, he kept reassuring himself, it would be to the west, and not by much.

After a moment Keefer spotted a perfect landing area ahead, just beyond a small farmhouse. "Hit the silk, Jack," he ordered, afraid that if they waited any longer they wouldn't need their parachutes.

"As far as I know," Jack continued, ignoring his friend, "neutral countries aren't interested in prisoners. If there is a camp like Dolly says, it must be for the Jerries. Everybody hates those bastards."

"Jump for chrissake, Jack!"

"With four or five hundred JU88s strafing London, a few are bound to get lost."

"Yeah, like us. Now jump, Jack, and save a Guinness for me!"

Keefer finally pushed his friend out, with Jack singing out the first few notes of their favorite song – "It's a Long Way to Tipperary" – as he dropped from view. "It's a Long Way to Tipperary," an old World War I marching song that the two sang together whenever the occasion suited, whenever they needed a lift. Like last call at the Windmill, or at Minsky's, or at the Rainbow room at the Regent, or while taking the fast train back from London, or flying low over the Humber River on their final approach, knowing they had survived. What style, thought Keefer, as the final notes echoed back beneath the fuselage, inviting the world to recognize such panache in the face of their first jumps ever!

Yet only Keefer remained, smiling at the thought that neither he nor Jack had any idea where Tipperary was.

Or that they had just flown over it.

While the exact aeronautical details of Keefer's feat may never be known, it can be safely said that with a distance of 440 miles from Frankfurt to his base at Elsham Wolds (the 103) and a cruising speed in a Wellington at light load of 140 MPH (Indicated Air Speed), the flight home should have taken three hours and ten minutes in still air – considerably longer than today's commercial flights. Or, as he often said, you can drive to or from Frankfurt faster than that, especially if you drive like the Europeans do. Anticipating a light headwind, therefore, Keefer and his crew should have reached their squadron at 05:40, having turned for home at 02:30, soon after dropping their load. So while Keefer going back to free his rear gunner may have been a contributing factor to their overshoot, and certainly made for a good story (which was just like him), the primary cause was undoubtedly the wind, specifically an unexpected 75 MPH tail wind that sent them an extra 200 miles over the Irish Sea, and a 50 MPH cross wind that caused the 150 mile shift to the south. This left a resultant reading of 90 MPH from 073, information which was no longer available to them, thanks to damage from flak. Many thanks to Walter Thompson, DFC and Bar, of Langley, BC, (author of *Lancaster to Berlin*, Goodall Publications, London, 1985, an excellent account of the RAF bombing of Berlin in '43) for those observations.

2

SLAINTE, AND WELCOME TO COUNTY CLAIRE

In fact, the former football star barely survived his first parachute jump. He landed in a peat bog, and as he did his right knee locked in the mud and his body wrenched forward, with needles of pain shooting up his leg. The chute collapsed around him, and for a moment he thought he would drown – the fate of more than a few RAF flyers who landed in bogs during the war. He managed to escape by crawling on his belly to a nearby haystack, peat seeping into his flying boots and pants, under his bomber jacket, into his sweater and shirt, and under his helmet, leaving only his nose uncovered. Given the smell, he may well have preferred the alternative.

Brushing the mud and debris from his face and uniform, Keefer stood up. Reaching down, he felt for the displaced cartilage above his right knee – another old football injury – and clicked it back into place. It would be some time before he could walk, let alone dodge onrushing Irish soldiers on his broken field dash to Belfast.

After a short rest and a quick check of his K rations and escape kit, Keefer spotted a small cottage in the distance. He limped off in its direction. As he reached the top, a glint of sunshine rebounded off the building's thatched wheat roof, leaving a bright yellow glow. An older man emerged from the cottage to greet him. The man was a turf-cutter and the owner of the bog in which Keefer had landed. Keefer could only describe him later as shabbily dressed, although the man wore a three-piece suit, buttoned to the waist, with tweed trousers and a white shirt, without collar or tie. The man also wore an oval cap known as a trilby, a battered one that would have been new twenty years ago. To Keefer, it seemed like an odd way for a farmer to dress; maybe the man had just returned from the market! He didn't realize that most Irish farmers dressed that way.

He also noted tools and farm implements strewn about, and smoke coming from the top of the thatched roof. And another smell – one that strikes all first-time visitors to that part of the island, whether they land by parachute or not – the smell of burning peat.

"Looks loike ye be takin' a dip in da bog," the man said, nodding. Keefer nodded back, shivering from the cold.

A woman poked her head outside. "Jackie, come inside, 'tis bloody cold ..." She stopped as she saw the sight before her. "Wisha," she exclaimed, clearing the way for the stranger to enter. "Looks loike ye need a scrub!"

The turf cutter cleaned his boots at the door. He had tied his pant legs at each ankle with hemp in order to work in the bog; as Keefer waited, he untied them, folding the two strings neatly into his right trouser pocket. Once inside, he offered Keefer a seat by the fire and a thick slab of bread, almost green in colour.

Keefer's spirits lifted as he limped inside.

"'Av ye thrown a name on m'y yet?" the man asked, in a greeting typical of that part of the island, where a familiar face was always assumed.

Keefer waited, not sure what the man meant or how to respond. Children, meanwhile, were climbing down a ladder from a loft above, five at last count.

"All roight den," the man continued affably. "My name is Jack O'Mallye of Knocknahila. Dis is my family: three sons, Michael, Edward and Paddy, and two daughters, Liliane and Catherine, and my wife, Brigid. And ye're wailcome to stay as long as ye loike. When we 'av a loikin' to someone, den we do it."

That sounded promising.

"Arrah, a wee bit lost are ye," continued Jack O'Mallye of Knocknahila. "Ye're on da town land of Knocknihila, near Drummin, two moiles nort-west 'a Kilmihil."

"Near the coast," Keefer asked with a nod, encouraging the turf cutter to provide a little more detail.

"Da very t'ing, da very t'ing," the man replied.

"Am I far from the North?"

"Da nort?" the man asked. "Ye mean Milltown Malbay?"

Keefer took his eyes off the man for a moment. No, I don't mean Milltown Malbay, you idiot, I mean Belfast – though he would scarcely have said such a thing, having just been invited into the man's home. He recalled a look of suspicion on the woman's face at that point. He then looked over at the children, who seemed to be frowning at him as well.

Mr O'Mallye got up and went over to a small wooden desk by the door. From the drawer, he pulled out a tattered map and sat down beside the Canadian "Ye're roight here, lad, jist besoide this little dot Knocknihila, ye see dat? A wee bit west of Kilmihil in da county Clare." Obviously, fond of giving directions, the man was pointing to the left side of a map of Ireland, on the western coast, about two-thirds of the way down.

That concerned him. "Am I far from the North?" he asked again, a little louder this time.

The Irish farmer looked at him closely for the first time. The man may not have heard the crash but, unlike some of his neighbours, Mr O'Mallye knew all about the war, England's war, since he had a wireless, one of the few in that area, in the corner of the room beside a bookshelf. Keefer, much to his chagrin, had just spotted it, and was busy planning his next move.

Once more Mr O'Mallye said nothing, studying the pilot officer's uniform, including the RAF insignia and flying dress, or at least as much of it as was visible beneath the peat. The man was obviously torn, as many Irish are, between Irish hospitality on the one hand, and distrust of the English on the other. And wherever the stranger had come from, it certainly wasn't Germany.

"That ye b'y," said Mr O'Mallye finally, moving his finger up the map to a red line at the top-right corner. "About 250 moiles to the 'sout, oi'd say."

Keefer felt a sudden pang of disappointment.

"T'is wan soft mornin' then," Mr O'Mallye said, after another moment of tense silence. "Clare and broight, and a plaisure for this late in the year, don't ye t'ink?"

Keefer just sat there.

"Looks loike ye've come a fair distance," he continued, closing in for the kill.

Keefer nodded.

"From where exactly?"

Keefer knew that Jack O'Mallye of Knocknahila wasn't just making conversation. He knew it would only be a matter of time before the police arrived.

Actually, it wasn't the police (or garda as they're known in the Republic), who arrived to arrest him. It wasn't the army either. It was the Local Defence Force, or LDF, a civilian organization set up to defend the Republic from a German, or English, invasion during the war. They looked like army officers, mind you, as long as you didn't look at them too closely. And if you spent any time with them, as Keefer was about to, then you would certainly know the difference.

The two identified themselves as Captain Michael Moronay and Corporal Kieran Lillis, and smiled as they greeted the pilot officer in his chair by the fire. They appeared to be about his age, in their mid-twenties. They would have had day jobs if there had been day jobs to be had but there weren't – not since the embargo of 1939 when Britain had taken steps to shut off Irish ports in retaliation for Irish neutrality. They were dressed in green military garb, but their coats were shorter and cropped at the waist. The rest of their uniforms were missing. And, rather than rifles, they carried wooden canes that looked sort of like rifles. And they arrived in a taxi, a shiny new black Addler, which they proudly showed Keefer as they ushered him outside into the warm sun.

If only Jack was here, thought Keefer. He'd make a great story out of this.

Captain Moronay and Corporal Lillis politely asked their new prisoner to get into the back of the taxi. Sarcastically, Keefer thanked Jack O'Mallye of Knocknihila for the bread and tea, and said goodby. The turf cutter nodded that imperceptible Irish nod, and then shook the Canadian's hand. "Slean leath," he said, apologetically.

Keefer was probably under arrest as he drove off, at least from a legal point of view, but the LDF officers never touched or handcuffed him. They did ask him for his gun, however, which Captain Moronay quickly slipped under his jacket.

When they weren't giving directions to the confused driver, the two LDF officers were asking Keefer questions about his home and crew and smiling a lot, saying things like "wailcome to the Free State" (still a familiar greeting at that time) "tail us about yerself," (an intelligence technique) and "looks loike ye be takin a dip in da bog," (an observation made by most people he met that day.)

Naturally, his arrival was big news, since they'd never had a crash in county Clare before.[3] They also told him that the other members of his crew had been found, including a red-headed Canadian. At least the two of them were in the same pickle, thought Keefer.

It seemed odd that the taxi suddenly stopped only a few minutes later, since there had only been time to pass a couple of farms and crossroads. The mystery was explained when they directed him into a white building and seated him at the table closest to the door. He steadied himself – this would be where they'd grill him! He noted writing in chalk on a black board to the table's right. A man came over, greeted the two LDF officers, whom he obviously knew, and said something that Keefer didn't understand but would come to recognize as Irish Gaelic, a language frequently spoken in Clare at that time. A moment later a bottle of Irish whiskey arrived – "Paddy," read the simple white label – together with three empty glasses. Captain Moronay filled all three to the brim and slid one over to the Canadian.

The Canadian slid it back.

"My goodness, that's not fer us," the Captain said with a sly look. "What koind of hosts 'ud we be?"

"Don't worry lad, it goes down noice and aizy," the Corporal said.

"Welcome to Erin!" the Captain declared.

"May ye live all the days of yer loife," said the Corporal.

The Canadian, still chilled from the long flight and with a knee that was swelling rapidly, could see no immediate harm in it. He reached for the glass.

"Throw that 'cross yer chest!" Captain Moronay ordered.

The Canadian hoped that the Captain didn't mean it literally. He raised the glass to his lips. The two LDF officers did the same.

"Slainte," they said.

3 At least not according to the Warplane Research Group of Ireland, who should know. (The Irish have research groups or associations for just about everything.) There were however, several war-time crashes and force-landings in Clare after Keefer's.

The O'Mallye girl describing the moment to Tomsy O'Sullivan fifty two years later.

Corporal Lillis and Tomsy O'Sullivan, 1993.
Captain Moronay passed away in the late 1960s.

"Slainte," he said, repeating their pronunciation.

They did that three or four times.

"Dat's a nice gun ye have, lad," Moronay said, after a while, patting his hip protectively. (If I could only get my gun back, thought Keefer, I'd put an end to this nonsense.) "We shure cud use guns loike dat here. All dey ever give us is wooden canes. How in heaven's name can we defend Erin with wooden canes?"

"Wisha," added Lillis. "Do dey honestly t'ink that the Germans won't land if we're standing on the beach waiving wooden canes?"

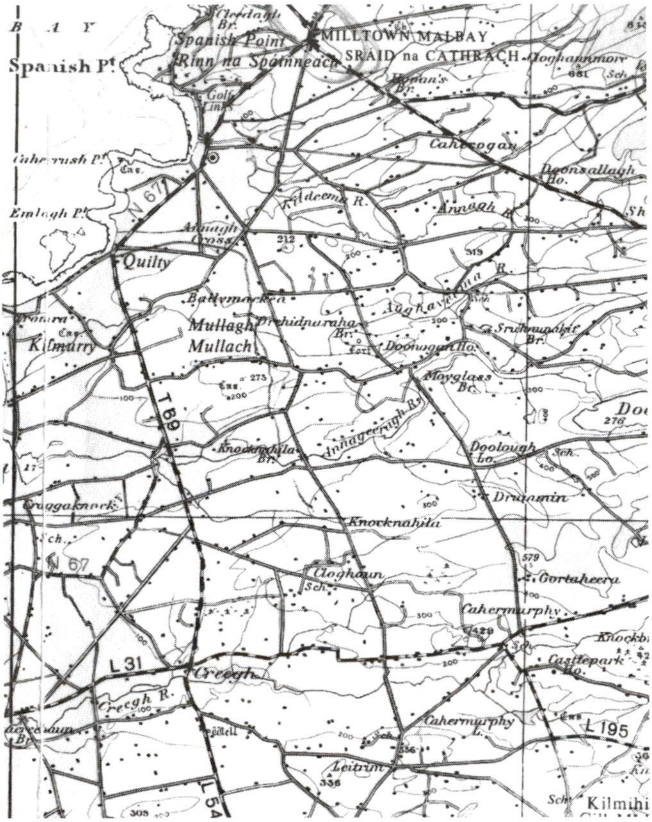

A map of northern Clare, including Knocknahila, a small townland, as the Irish call them (middle), Milltown Malbay (top), Kilmihil (bottom right) and Quilty (just below Spanish Pt., top left.)

"Would you loike a pint den before we head out," asked the Captain, "or more whiskey?" It was 11:00 hours.

On the way out he noticed a sign above the door with the words, "Meaney's Pub."

Back on the road another sign suggested that they were about to leave a small hamlet called Creegh, to the south of Knocknihila, and head north, on what he hoped was the road to Belfast. A few minutes later, however, they stopped at another crossroads where the sign on a corner grocer's shutter suggested that they had arrived in a place called Mullagh – unless a Mullagh was a person or a thing – where they visited O'Brien's Pub. There they met two other LDF officers, similarly dressed and leaning on their wooden rifles. They ordered a round of Guinness, and Keefer was commanded, once more, to drink.

Next they drove either north or south – he lost track – to a place called Inagh (Leyden's Pub), and then finally east or west, depending on the previous location, to another town and another pub, although he couldn't recall either name. Soon they were traveling in a procession of cars, bicycles, carts, cattle, horses, you name it, (as if he were an important diplomat, he used to say, or even a head of state!), sometimes stopping to talk to people on the side of the road with animated gestures in his direction, at other times backtracking to the last crossroads, or getting out of their cars and looking at small maps that he couldn't read, or just plain driving around in circles, all in the relaxed confusion that strikes anyone in the West of Ireland who makes the mistake of trying to get from one place to another.

Still there fifty-two years later!

"Jaysus, Kelly," yelled Corporal Lillis, slamming his Guinness down on the bar. They had been drinking at Casey's Pub, also known as the Quilty Tavern, for an hour or so waiting for the Irish Army to show up. "The army would give us more guns if you learned how to clean 'em properly." The corporal was upset with his neighbour Sean Kelly, seated at the bar to Keefer's right, who felt that the government in Dublin wasn't doing enough to help the LDF, or at least their chapter of the LDF. "Ye gotta open da chambers to git da bullets out first. T'is no good to jist fire da t'ings into da door, for goodness sake. That won't clean a gun."

"N'deed it wont, n'deed it wont," added another. By now the crowd had swelled considereably.

"Dat's a load of bullocks and ye know it." This fellow Kelly was no pushover. "De only reason dey won't give us the guns is 'cause the bloody English won't give dem the guns. We're a poor country, lads." Everyone nodded in agreement at this. "And oi'll take da Germans any day. At least they helped us." Kelly was referring to financial and military assistance provided by Germany, to the IRA in particular, during the last war. He might also have mentioned the Spaniards, who had landed at the same spot three hundred years earlier, during the War of the Roses.

"Jaysus, Kelly, you can't be serious." Lillis looked quickly at Keefer, embarrassed in the Canadian's presence. "The bloody Germans would be worse than the English. What have they done to Europe? Slattering da jews. Look what dey did to the Poles. We're a small country, and if it wasn't for the English ..."

"Lads," interrupted Captain Moronay, who could see no good coming from an extended debate on this subject. "Where should we be taking our prisoner next?"

There was silence until someone yelled out "Has anybody talked to the army?" and then more silence, and then laughter all around, since their mission, which they had bravely accepted, had been to entertain him until the army arrived.[4] And then someone named Tomsy O'Sullivan bought another round. Keefer, meanwhile, raised his increasingly heavy head to study the innkeeper. The man acknowledged the last order with the same imperceptible nod as the turf cutter, then carefully guided the black beer into the numerous glasses before him, pausing to let it settle before proceeding further. It involved a three stage process and Keefer was impressed. There were army surgeons with less concentration than this guy.

"Lads wait 'til I tail ye dis," the man named Tomsy O'Sullivan announced suddenly from his seat, two stools down on Keefer's right. (A ritual of any Irish pub, thought Keefer, a few jokes.) "There wuz two Englishmen in a rowboat in the middle of the ocean, or maybe I should sey two Canadians," he said, smiling over at the pilot officer. "One of them is rowin' loike hell, and the other's in the bow. Suddenly, an elegant ocean liner comes into view. Da one in the bow yells at the tap of his lungs, 'is this the Atlantic or

4 Keefer's grand welcome that day in county Clare was not atypical. Of the 112 RAF landings and crashes during the war, over half of those detained were taken into custody by the local guard first, and many were entertained in similar fashion.

Pacific ocean?' Da captain yells back, 'Pacific,' and den the Canadian who wasn't rowin' turns to his friend and sez: 'Jaysus, I told you not to row so fast.'"

The man named Tomsy O'Sullivan then let out a roar, slamming his beer down on the bar. He looked over at Keefer.

Keefer – assuming correctly that he had been the butt of the joke – thought he should reciprocate. He knew one, with a few amendments, that might work.

"At one time there was a drill sergeant in the Local Defence Force," he began, once more seeing no harm in it. "His name was Kelly, Sean Kelly." He already had a few smiles and nods. "This Kelly was put in charge of a group of young recruits. He gave them a right dress command, but try as he might he couldn't get them to form a straight line. So he shouted ..."

Keefer paused dramatically to prepare the stage Irish accent that he had picked up at the Hunter's Horn on Ste-Catherine Street in Montreal, one of his favourite pubs. He and Jack had spent some time there three years earlier, after McGill's football victory on the night of the big riot, leading him to wonder, for a moment, how Jack's interrogation was going.

"What's da madder wid ye? Can't ye loine up straight for the sweet love of Mary?"

His audience began laughing now, and he hadn't even arrived at the punch line.

"But still the officers in the Local Defence Corps couldn't get it right ..."

"And den what," asked Tomsy O'Sullivan, impatiently, "and den what?"

"Finally, the sergeant issues an order: 'Dat loine is as crooked as a cork screw. All of yez fall out and tak' a look at it.'"

The place went wild.

It remains a mystery why the LDF officers, after the army's arrival in Quilty, loaded Keefer back into the taxi and drove him north, if only briefly. They could have walked him there in five minutes, Casey's being a stone's throw from the Garda station where the transfer was to be made. Keefer, whose plan, had been to play along with his captors knowing that good will was his only hope, couldn't shed much light on the issue. He didn't remember much

Quilty Bay

after that second round at Casey's since his popularity soared with the telling of the first joke such that he was obliged to tell a second and then a third. Indeed, it was another few hours before they allowed him to leave. He recalled talking to Captain Moronay between pints as they stood in the bright sunshine in Quilty overlooking the bay. He was staring out at the ocean, he recalled, limping along a horse-shoe-shaped stone wall that separated the beach from the village, explaining his plight. He thought that maybe he could hop a boat – though God knows how he'd manage to accomplish that with all the German subs and drunken Irishmen around. He remembered what Moronay had said about his being Canadian, thinking that the Irish might release him or drive him to the border because of his nationality. He wasn't, after all, English, and he

The Garda station in Quilty – the slammer!

knew enough history to know that that mattered. But he wasn't about to leave his crew. He remembered sitting in the back seat of the taxi while Moronay had an animated discussion in Irish Gaelic with an LDF officer in the front but, given his condition, even if he had known Irish Gaelic, the nuances would have escaped him. They might well have considered releasing him, for not every one he met that day was his enemy, that much he knew. But all that really mattered in the end was that they didn't.

When Keefer finally arrived at the Garda station, it was obvious that things had changed. He attempted to shake Moronay and Lillis's hands but they retreated quickly in the presence of Irish Army NCOs and officers who had arrived from Ennis. An older man, in a crisp uniform with a German-style coal-scuttle helmet, arrived. He appeared to be in charge, ordering the civilians away. He directed one of the NCOs dressed in similar garb to usher Keefer inside, which the NCO did gladly, thrusting the barrel of his rifle hard into the Canadian's back. Keefer was led down some stairs, directly to the door of a plain white room, where two sergeants stood guard. He looked inside.

Jack was seated at a table with his back to the door, bent over at the waist with his upper body prostrate. He was motionless, and for a moment Keefer thought he was dead. Until he heard the snoring.

After he had parachuted to safety Jack Calder had sat in the only dry spot of the bog in which he had landed, eating his RAF-issued dry chocolate and thinking about what Dalton had said. If Dolly was right, as Jack suspected, then he wanted very much to escape to the north, despite what he had said earlier. The notion of being locked up briefly might be appealing, especially in a brewery or, as a reporter, in a place as incongruous as Ireland, a fellow member state of the British commonwealth and eventually, the only English-speaking country in the world to remain neutral. That would be a story in itself. But as an airman, he could think of nothing worse than sitting about, brooding, in the aftermath of this botched mission.

After the sun had risen, Jack escaped the bog and began looking for a hiding place for his parachute. He found a small copse at the base of an oak tree and ripped the badges and insignia from his flying dress – one step ahead of Keefer on that score. He saw a cottage and headed in that direction until a child spotted him climbing a stone wall.

"O, momma," she cried. "Lookit the man!"

A moment later the family came out of the cottage, only to see Jack running at a gallop down the road. He was headed towards the coast, he later recalled,[5] passing cattle drovers and donkey carts along the way. He was probably running down the same road north of Quilty (a stone's throw in the other direction from Casey's) that Keefer had driven up at least a few times earlier that day, leaving them to wonder later why they never met. Reaching the coast, Jack then saw what he had been looking for, railway tracks. He sat down on the side of the embankment and waited patiently for the next train.

"Where are ye goin' and where are ye comin' from?" said a voice from above, an hour later. Jack turned around to see a gardai resting on his bicycle just above the railway embankment.

"I'm from the sout and I'm going nort'," Jack replied casually, using the exact words which he had intended to use, for he had also practiced his brogue at the Hunter's Horn in Montreal. He wasn't sure how far south, or how far north, but it sounded good.

He had a distant aunt somewhere in the south of Ireland, he

5 In an article entitled *I Flew into Trouble*, eventually published by *Maclean's* magazine on 15 August 1942. For the impatient reader that article appears in its entirety at 186–7. If you can't believe what you read in *Maclean's*, my dad used to say, what can you believe?

remembered thinking, though he wasn't sure which county. Maybe he should look her up.

"Wail, d'em as goes nort' always comes into our barracks for a nice cup of tea," the gardai replied just as casually, dismounting from his bicycle.

Jack felt the gardai frisking him with his eyes, so he surrendered his Browning rather than risk an incident. They walked a mile or so together to the Garda station.

"We heard ye wuz comin," said the man, Sergeant Doherty, the one and only gardai in Quilty. "Yer friends will be along in a twinkle."

After they arrived in Quilty, the gardai invited Jack into the back of the detachment, where he and his wife lived, to await the army. While his wife prepared an Irish fry of bacon, eggs, beans, and toast, Sgt Doherty showed Jack photographs of their children and grandchildren, of their trip to England before the war, and of their cats. Before long Jack had begun to feel as if he was indeed visiting his aunt and uncle, wherever they might be, rather than the Irish garda.

Sergeant Doherty had been a member of the garda for many years, he explained, and the only officer in Quilty for the past dozen. He said it would take the army at least a few hours to reach Quilty, noting that a posse of Local Defence Force officers and part-time garda had been sent out to look for anyone from the plane. "Da whole county is on about it," he said. After breakfast, Doherty's wife offered Jack a bath, which Jack gladly accepted. She then whisked Jack's clothes clean as her husband went to the cupboard and got out a bottle of Jamieson's, a smoother blend of Irish whiskey than Paddy's. By the time Jack had emerged from the bath his clothes were clean and Sergeant Doherty was waiting in a small dining room for his prisoner with the bottle and two clean glasses. Mrs Doherty sat down with them as well, although she preferred tea.

As time passed, they talked about many things. Sergeant Doherty, who was well-informed on current events, explained that not everyone in that part of Ireland distrusted England, but everyone supported neutrality. This was because county Clare had always been pro-de Valera, having elected the Irish leader to the Dail, or the Irish parliament, in each election since 1917. They didn't hate the English, they just distrusted them, which had always been de Valera's view. If the Irish joined England, he noted, Ireland would be vulnerable to attack, a poor country with little in the way of weapons to defend themselves. If they joined the Axis, England would annihilate them. Either way, the Irish couldn't, and wouldn't, permit the island to be used as a landing strip. When Jack raised the issue of Irish ports and German U-boats, thinking not only of what Dalton had said but of what he had read in his eighteen months as an editor with Canada's national wire service, Sergeant Doherty pointed out that these had been Irish ports to begin with, not English, with names like Berehaven, Cobh, and Lough Swilly, names which Jack might not have heard of but certainly found easier to pronounce than the Welsh ones he had seen on his map a few hours earlier. When de Valera, or de V as he was known, had negotiated their return from Chamberlain (before Churchill took over in 1939), it was because of a long-standing treaty obligation under the agreement of 1921, the Partition Agreement, not because of Irish spite, or deceit, or unfair capitalization on Chamberlain's weaknesses – views widely held in England at the time.

Jack, former reporter, airman, and loyal servant of the Crown, listened politely. He then started arguing. He also asked about the camp Dolly had mentioned. Sergeant Doherty replied that he had read of a camp being built up north, near Dublin, an internment camp, for both English and German crews, but he didn't know what the government intended to do with them. If the army thought they had been on a training mission, rather than operations, they might be released – he had heard of that happening. Or if they were Canadian, since at least one of their crew members was Canadian (Jack had said nothing about the others, including Keefer) or, even better, Irish Canadian, (Jack having mentioned his aunt) or, even better than that, Irish American (interesting, thought Jack, remembering his distant American relatives in Boston). After all was said and done, however, Jack was grateful for the hospitality, and soon the effects of the sleepless night, the warm bath, and the Jamieson's took hold.

The senior army officer who had ushered Keefer into the garda detachment sat down and introduced himself. He was Major O'Connell and he had had just come from the Irish Army's southern command headquarters in Limerick. After waking Jack up with a push, he began pressing him for details of ditching the plane. Jack kept silent, confused from sleep. The officer left for a moment, leav-

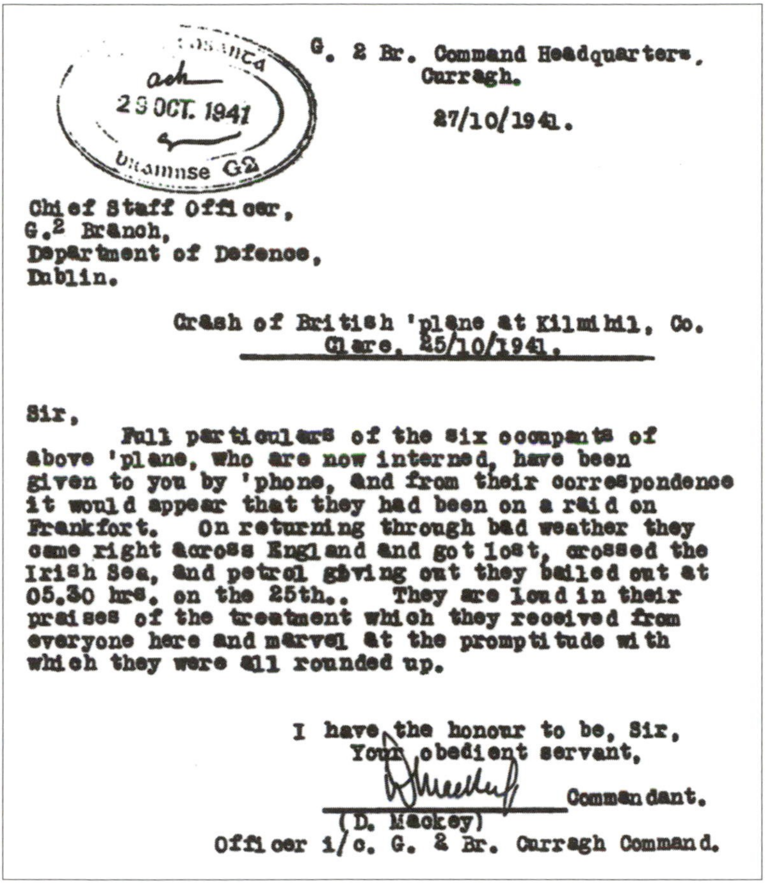

28 OCT. 1941
Brainse G2

G. 2 Br. Command Headquarters,
Curragh.

27/10/1941.

Chief Staff Officer,
G.2 Branch,
Department of Defence,
Dublin.

Crash of British 'plane at Kilmihil, Co.
Clare. 25/10/1941.

Sir,

Full particulars of the six occupants of above 'plane, who are now interned, have been given to you by 'phone, and from their correspondence it would appear that they had been on a raid on Frankfort. On returning through bad weather they came right across England and got lost, crossed the Irish Sea, and petrol giving out they bailed out at 05.30 hrs. on the 25th.. They are loud in their praises of the treatment which they received from everyone here and marvel at the promptitude with which they were all rounded up.

I have the honour to be, Sir,
Your obedient servant,
Commandant.
(D. Mackey)
Officer i/c. G. 2 Br. Curragh Command.

From the Irish Military Archives

ing the two Canadians, and the bottle of Jamieson's, to reflect.

"Jesus, don't touch that stuff," Keefer said, trying to focus on the half-empty bottle. "It goes down like fire and hits bottom like a sledge slammer."

"Bobby, you're pissed," but before Jack could say anything more, the major returned, together with the two sergeants who had been waiting outside. He studied Keefer for the first time and sat down. The two sergeants remained standing by the door.

"As I was saying, I'm Major O'Connel from Limerick," he announced. "I apologize for taking so long. Now let's get down to business. Where have you come from?"

There was no answer.

"Come, come, what's the secret? You announced your arrival on the wireless across three counties for goodness sake!"

So the radio was working. "You heard our Mayday and you didn't answer it," asked Keefer.

"That's correct.[6] The two of you must have stayed for a nip, since your friends landed back in Kilmihil." The major smiled. "So you came back from Germany on a bombing mission, did you? That's what we found at the crash sight. We'll be taking you there presently."

So much for the ditch, thought Keefer. Obviously, he had misread the wind again when he had reset the tabs for the last time.

"No sor, not at t'all, not at t'all. We've come from Ireland, initially," Jack began slowly, following the advice of the kindly Sergeant Doherty. "We live in Canada but we 'ave American cousins."

"Shut up Jack."

"And where moight ye be headed," the officer asked, ignoring Keefer and imitating Jack's stage Irish.

"Why, Northern Ireland, of coourrrse."

The officer was smiling. "I see."

"I'm jist leveling wid ye sor." Jack was adjusting his brogue, which unfortunately had a tendency to sound Scottish whenever he had been drinking. "My wee frriend and I both live in Canada but we were borrrn in Dublin before our family moved to Boston when we were wee lads."

"Then why are you with the RAF?"

"We're with the RAF all rrroight, but we got to tinkin' that we should visit the old country on our next Rrr & Rrr."

The officer began laughing.

"Seriously sirr, ever since our families come to North America we've been greatly interrrested in da Gaelic revival back home. And dat police strike in Boston? By Jaysus 'dos boys deserrrve more money, don't ye t'ink! Why me mum wrote me just last week, she sez, Jackie me boy, de O'Reilly's and de O'Flaherty's, yerr good cousins from Boston, they tail us that ..."

The officer slammed his hand down, hard, on the table.

"So what is yer name boy?"

"Why Jackie, sir. Jackie *Fitz*calder."

"And you, then?" The officer looked over towards Keefer.

6 While the Irish habit of ignoring Maydays during the war infuriated the English government and cost a number of RAF lives in the process, to do otherwise would have jeopardized their neutrality, or so the argument went.

"His name is *O'K*eefer sir," Jack replied. "*Babby O'K*eefer."

"I said shut up Jack!"

"Is he your commanding officer, Mr Jackie Fitzcalder?" Neither spoke. "And you're part of this Gaelic revival, too, are you, young Babby O'Keefer?"

Keefer glared at his navigator.

Major O'Donnel turned back to the other Canadian. "Which county are you from then, Mr Jackie FitzCalder?"

"Hmm ... Belfast sir," Jack replied.

The officer studied the two Canadians for a moment, amusing himself. "That's a grand way to dress, young Jackie FitzCalder, visiting your dear family in the county Belfast, in the north. Is the RAF flying tunic part of the Gaelic revival as well? Interesting way to arrive at a family reunion – by parachute."

Jack suddenly decided to renounce his Canadian-American-Irish heritage.

"Actually sir, we've been lying to you," he said, not exactly surprising his interrogator. "We're nervous, that's all. The truth is that a half dozen of us have landed in a training exercise – by mistake – and we'll be picked up by the RAF in a short while."

"The RAF has training sessions in Erin, now does it?" The army officer was grinning widely. "Well then, why don't we just sit and wait for the defenders of the Empire to show up. They can tell us what to do next."

Keefer wanted to join in, to tell the officer that in fact they were Canadian, that they were just volunteers, that they had nothing against Ireland, nor Ireland against them, that they meant no offence. That they were merely fighting someone else's war.

But he couldn't do it. He still didn't really understand the historical thing, but he wouldn't have done it anyway. It would have been wrong. So he let Jack do the talking.

At 15:30 Keefer and his navigator were driven by lorry to the scene of the crash. There they encountered garda from Ennis, cavalry officers who had arrived by motorcycle from Limerick, and numerous other Irish army officers, real ones this time, with full-length green coats, real pants, and real guns. The soldiers had cordoned off the area and were stripping the Wellington, ripping the canvas from its aluminum frame like skin from an animal's carcass. The rear turret

Virtue appears fourth from left. Courtesy Warplane Research Group of Ireland, M. Gleeson and C. Vance

had become separated from the fuselage and Virtue had begun posing in front of it like a tourist, with the Irish army and cavalry on either side of him. "Acting the fool," as Brownie later described it. At one point, Virtue turned on the officers who were guarding him, scuffling with them, challenging their manhood, accusing them of seizing the guns in order to give them to the Germans, of not having anything better to do with their time. He then tried to pawn his turtle-neck sweater in exchange for his freedom, but his captors would have none of it. A major confrontation would have broken out had Keefer not ordered him to be quiet.

The soldiers seized some 1,500 rounds of live ammunition plus two Browning guns from the two turrets. And everything they seized and observed – including the bullet holes in the fuselage – led Major O'Sullivan to conclude, following consultations with Parkgate in Dublin, that the Wellington had crashed on its return from a bombing mission in Europe. An operational flight in other words, meaning that the entire crew would be detained.

The Canadians, who now knew that they were in trouble, were given permission to wait outside in the warm sunshine with the rest of their crew for the arrival of another lorry which would take them to the Army barracks in Limerick. Ten minutes later the prisoners were handcuffed and driven from the scene.

Throughout the two and a half-hour trip to Limerick, the five ton, canvas-covered lorry threw the men around like mice in a cage. The route took them east through the Clare countryside, then south-east,

Sgt Doherty, the taller of the two police officers, appears left and again bottom right.

Immediately below, an Irish NCO poses with Virtue in front of his turret. Both photos courtesy of the Warplane Research Group of Ireland, M. Gleeson and C. Vance.

The Wellington after the crash. Cyris Vance is shown above, far left, at age 7. His mother, who took the photograph, has since died. Courtesy Warplane Research Group of Ireland, M. Gleeson, C. Vance.

and then north along the estuaries of the river Shannon. En route they passed men in suits tilling their fields and women washing in the river with children playing at their sides. Had it not been for the company the setting would have been pleasant enough. The Shannon Estuary was not unlike the Ottawa valley that both Keefer and Calder knew so well, the misty, gentle slopes of the delta's banks, the birch and willow trees, and above all, the relaxed faces and seemingly friendly dispositions of people who hadn't been living in fear of night raids for the past two years. They also saw strange-looking carriages travelling in caravans across the horizon. These, one of the guards explained, were owned by gypsies, known as tinkers, who roamed from county to county claiming to be dispossessed Irish gentry. Although disliked in some quarters, tinkers were usually allowed to help themselves to whatever they could find in the way of food. It may seem strange, the guard agreed, that one would allow these nomads to steal food given the poverty of the island, but Ireland had its own pace, its own sense of time.

"No kidding," said Virtue. "Don't you Irish bastards know there's a war going on?"

"We've never had much to do with Europe, lad – that's your concern," replied the guard who had taken Virtue on earlier.

"Yes, well, it's everyone's concern. And where are all your bloody cars? And the people? I thought you people were part of the modern world!"

"Maybe not of your world, lad."

"World – this place is like another planet," snapped the Englishman, echoing in part the sentiments of the others, including the two Canadians: it was one thing to parachute into the wrong country, it was quite another to land in the wrong century.

"What is this nonsense," continued Virtue, "You've got officers combing the countryside for us and you won't lend a bloody hand to beat Jerry. How do you justify that? Come on coward, answer that!" The tail gunner then lurched hopelessly at the guard, who repelled him with a swift cuff from the butt of his gun.

In Limerick the six detainees, still handcuffed, were placed in two holding cells by Major O'Connell, who had arrived from Clare ahead of them. He released them at about 20:00 hours, allowed them to eat in the mess after the regular troops had finished their evening meal, and even joined them, but refused to answer any specific questions. He explained that they would be taken to the Irish Army's military base south of Dublin in a few hours and kept at a makeshift camp there until further notice. "You'll be guests of the Curragh, lads," he said, "but they'll show you a good time!" Major O'Connel wouldn't provide more details about the camp, referring questions on the subject to the camp commandant, a man named McNally. But he indicated that their country's diplomatic representative in Eire would be contacting them.

From Limerick they headed north, by-passing Tipperary without bursting into song, through Nenagh, Roscrea, Port Laoise, then finally the heart of the midlands and county Kildare. As the alcohol wore off, the reality sank in. Keefer sat quietly in a daze, neither angry nor pleased, neither defiant nor resigned. He looked over from time to time at his friend, and knew that Jack felt the same. Things could be worse, much worse, but certainly neither had expected this.

Still, he thought, if these Irishmen think they can stop us from our duly appointed task of saving the Empire, they've got another thing coming.

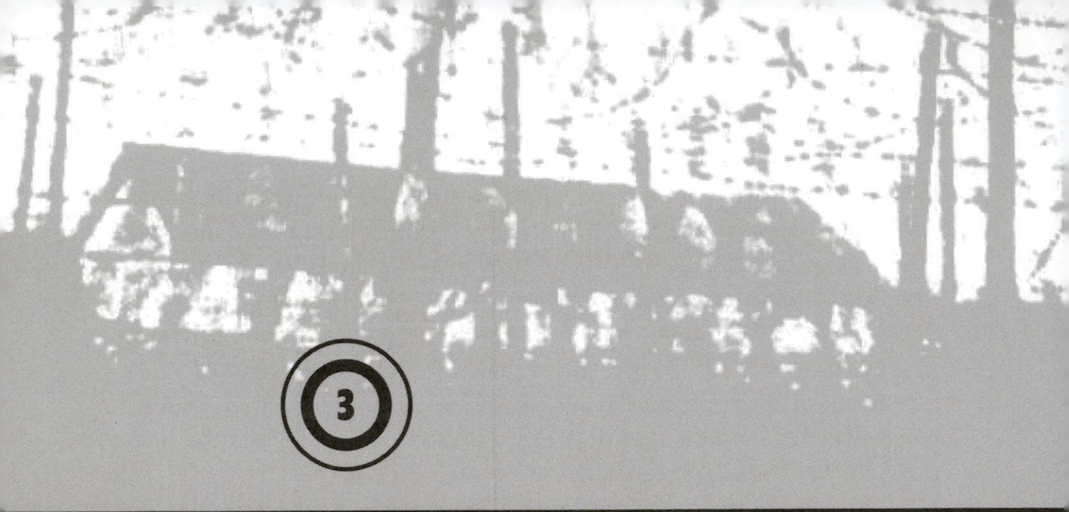

3

THE OFFICERS' BAR

It was high noon when Pilot Officer Bobby Keefer finally woke, shivering in his boots. A low-pressure system had moved in overnight and the clear, almost balmy, weather of county Clare had given way to more seasonal conditions. To make matters worse, the blanket that the army had provided was moth-eaten, and the small iron stove in the corner of his hut hadn't been filled with coal. So Keefer had huddled in a ball, on a thin one-inch mattress, cursing his luck.

When he did finally roll out of bed, he limped straight for the door and stepped outside. There he was met by another surprise: a small courtyard with other temporary wooden huts around it, each encircled by wire. He hadn't noticed the previous night, but some of the buildings were suspended two to three feet above the ground, elevated on concrete piers. It looked as if they had been built first and jacked up later, for the wooden staircases had been lag-bolted directly to the frames. It didn't take him long to realize why: someone must have tried to escape by digging a tunnel under one of the huts.

Keefer then turned to his right and saw an observation post and spotlight above him, explaining the glare of the previous night. Next to the post was a wall, ten feet high. He stepped a few feet closer. On the side of the wall was a single letter, painted in white, the letter G, for Germans. He'd been told that they'd be close, but he hadn't thought they'd be *that* close.

He continued turning to his right, but saw no sign of any other prisoners, either in the huts nearby or in the compound beyond the barbed wire. He did, however, see two Irish Army soldiers pacing back and forth in the observation post above him. They had dogs on a leash and wore the same dull green uniforms that he had seen in Clare and Limerick. He turned towards the main gate, which he saw was padlocked and guarded by two soldiers. They were staring at him, though no one said a word.

He completed the circle before returning to his hut. "You'll be guests of the Curragh for a while lads, but they'll show you a good time!"

"Welcome to the Curragh," he muttered in Jack's direction.

Jack turned his head from the wall, looked over at Keefer, and slithered off the bed. The two walked outside in search of the other RAF internees.

"Well, at least we outrank the bastards," said Jack nervously, looking up at the observation post above them as they circled the compound. Although the Irish guards they saw that day were only NCOs – wearing the international insignia of non-commissioned officers, one or two chevrons for corporal, and three for sergeant – they were armed and intimidating. They were older than most NCOs and several had been prisoners of the English during the Easter Rebellion some twenty-five years earlier. So the internees could expect little sympathy from them.

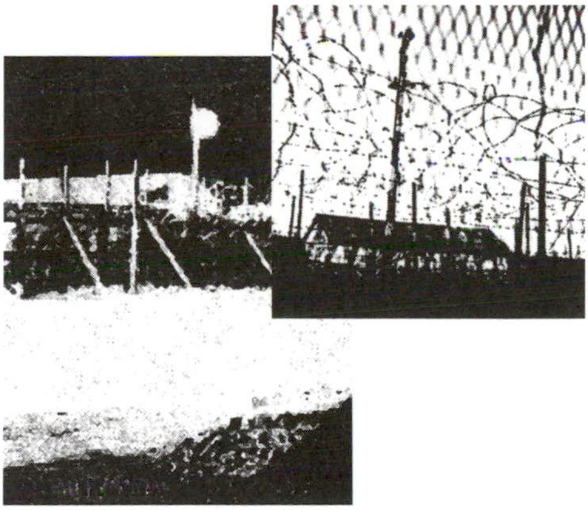

Several photos and sketches of the camp by Bruce Girdlestone, who arrived at camp in December. Courtesy Bruce Girdlestone

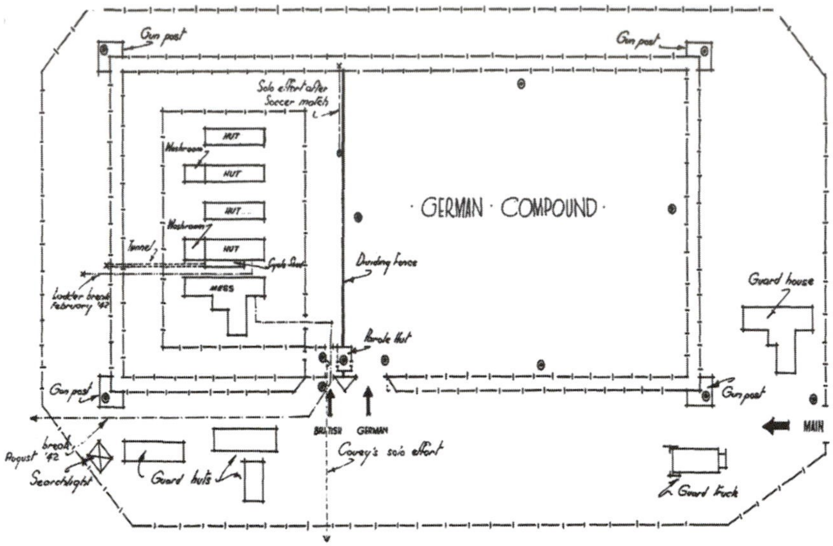

Reproduced by Bruce Girdlestone from a 1943 report on the camp prepared for the Department of External Affairs in Ottawa. The interior of the German compound was similar to the RAF compound with eight rather than five separate accommodation huts needed for their greater numbers, fifty-six compared to thirty-three Allied flyers as of late December 1941.

"Can you believe all the barbed wire ..." Keefer's voice trailed off as he studied the piles of wire, wrapped tightly in bales like wheat at harvest time, in some places up to twelve feet high, winding in and out of the various huts. "How in hell are we going to get through that?"

Jack shrugged, thinking the same thing. "Maybe there's a meeting going on somewhere," he added, for he hadn't seen any internees either.

Keefer didn't say anything this time, busy collecting his own thoughts. For one thing, he was worried how his mother in Ottawa would react to news of his capture. In the ordinary course he knew that she would receive a report from the Old Man at Elsham Wolds – *missing in operations* – which was guaranteed to upset her, thinking that she had now lost her youngest son after having already lost her husband, leaving only Eddie, Keefer's older brother, who wasn't much safer on active duty with the Royal Navy. Of course, she would celebrate when it was reported that he and Jack were still alive in a camp. But what would she think on hearing that the camp was in Ireland? Hopefully not that they'd flown there intentionally.

And what of the Old Man, and the others at Elsham Wolds, at the 103? What would they think? That he had let them down – which was how he felt too. He had hoped to become a flight lieutenant one day; after this snafu, he'd be lucky to remain an officer.

And, most important, how did he, Keefer, feel about the whole thing? Grateful that he wasn't standing in Germany right now, yet with all the guns and barbed-wire, was this place any different?

As they explored a small hut at the northern end of the compound, they received their first hint that it was. Off to the left in the foyer was a door with a sign mounted on it, a brass sign, with the words Officers' Bar engraved in a fancy script. They stopped and stared.

"What in hell is this," he asked his navigator, who had, after all, got them there.

"What do you think it is," replied Jack "It's a bloody bar ... Can't you read!"

"Do you think we should go inside?"

"Why not?" Jack replied. "We're officers, aren't we?"

For a while anyway, thought Keefer.

Interned in Eire

P/O. R. G. Keefer

Former McGill football and hockey star, and bomber pilot since last spring, P.O. "Bob" Keefer has been interned in Eire after a forced landing, his mother Mrs. F. K. Keefer, learned yesterday.

Ottawa Citizen, 30 October 1941

An early impression of their predicament, by Bruce Girdlestone

"Hello chaps, smashing of you to turn up like this." Keefer turned with a start, to see a tall man striding towards them. "Welcome to the officers' bar!"

At least, they'd found someone.

The man offered his hand, speaking quickly. "We were talking about you chaps last night: it'll be nice to hear the Canadian side of things around here. Calder, is it? So you nailed the *Gneiseneau*! Cheeri-bloody-o! Brest – awful place. Yes, yes. Keefer, the rugby star! Welcome, lads, welcome!"

The man, Flight Lieutenant Aubrey Covington, was tall, thin, and remarkably well dressed for a prisoner. He sported tweeds, flannel shirt, and wool tie. As they would soon discover, he had landed in Eire ten months earlier in December 1940 in a single engine Miles Master and was the Curragh's longest standing RAF internee. Hailing from Kingston-on-Thames, Covington had flown Spitfires during the Battle of Britain before that, and, like many Spitfire pilots, had picked up a case of the blinks along the way.

Flight Lt. Aubrey. Courtesy Bruce Girdlestone

"The bar officer." (The dog came later.) Courtesy Bruce Girdlestone

"Well, right to it, lads!" he announced, blinking incessantly as he spoke. "I am the bar officer and the first and most important function of the bar officer is to issue each new internee a set of bar keys." The Englishman paused to reach into the drawer of a desk beneath an open blue ledger. He passed the Canadians two keys each, one large and one small. "The large one opens the door, gentleman, and may be used at any time, day or night." He then lowered his voice to a whisper, blinking all the while. "And the small one, the drawer." He was looking intently at the cabinet beneath the bar. "The drawer, however, may only be used with discretion."

"Why," Jack asked. "What's in the drawer?"

"There, in that locked drawer," he replied suspensefully, "is the port!" – spitting the final word out as if describing a treasure of incalculable worth. "We call it the 'inner cabinet.'"

"'The inner cabinet' eh, sounds excellent," replied Jack, obviously pleased. A duty-free bar in a prison camp, mused the former reporter. I can get used to that.

"Where is everyone anyway?" asked Keefer, thinking much the same.

Covington ignored the question however, speaking quickly. "The rest of the bar is amply stocked, gents, so there is no need to worry about the inner cabinet. On this side is the Scotch, Gin, various types of Irish whiskey, and Guinness porter," he said, sweeping his left hand along the glass behind the bar, "and on this side is the French wine, sherry, cognac, and other liqueurs," sweeping his right hand along the other end of the bar, "and you really must choose since the day is upon us."

"You can't even get this stuff in England," Keefer remarked.

"Correct, my good man, and that's the way we like it. If the silly bastards insist on keeping us here, then we'll continue to insist on the minimum requirements for His Majesty's officers – would you not agree?"

"Who smuggles this stuff in here anyway?" Jack asked.

"The diplomats," Covington replied. "And I assure you that they fulfil no other useful purpose."

"What kind of diplomats?"

"British Representative, Canadian High Commissioner, French, Polish ambassadors ... that sort of thing."

"So how about that, Bobby? "High-Ranking Government

Officials Smuggle Liquor into Irish POW Camp." Not a bad start."

"Ah, yes ... very witty, Jack," said Covington, eyeing the Canadian with suspicion. "Once the newspaper man, always the newspaper man!"

Didn't take the man long to figure that one out, thought Keefer.

"The first thing a new officer asks when he arrives in the Irish Free State," continued Covington, "and more specifically this esteemed establishment, is 'how much?' I assume you two chaps will prove to be no exception. The scotch, gin, cognac, and sherry is five pence a shot, which is the duty-free price courtesy of the various embassies, and the Jamieson's and Guinness is free, courtesy of the Air Ministry and Sir Haldane Porter, a close personal friend of mine. Now what will it be?" He had leaned over the bar and was squinting at the two Canadians, waiting for their reply.

They looked at each other, shrugged, and sat down at the bar.

Keefer started searching aimlessly through his pockets. The bar officer looked at him with a knowing, discreet nod. He then reached behind him for the blue ledger that had been open on the bar and tossed it at Keefer, with the pen tied to the center leaf falling conveniently into the pilot officer's hand. "Given the auspicious occasion, I would heartily recommend a bottle of port." He looked furtively towards the locked drawer and then emerged a moment later with an unopened bottle of Madiera.

"Observer[7] buys the first round. Tradition."

So Jack did the honours, opening the bottle and filling three small glasses.

The bar officer nodded with approval and instructed Jack to enter the transaction into the book, under C for Calder.

"Chaps, I certainly cannot exaggerate the folly of the English government," Covington explained, sipping on his port, "when it failed Ireland – all because of the whining of a persistent and treacherous group of Republicans. I am greatly relieved that I was too young to witness the moment." Covington went on to describe how 2,000 troops loyal to King George VI had packed up and left the Curragh

7 Many flyers, especially permanent RAF members, referred to navigators on Wellingtons as observers rather than navigators. An observer in the six-man Wellington crew was responsible for aiming and dropping the bombs as well as navigating, hence the broader designation.

An aerial photograph of the Irish Army's main base on the Curragh, circa 1939. The internment camp was located in the foreground of the white tents barely visible in the top left of the photo. The golf course was just beyond that. (Thanks to Commandant Bill Gibson of the Irish Army.)

at 12:00 hours on 16 May 1922, leaving Lieutenant-General O'Connell of the Irish Army in charge of raising the tricolour. "There," he gestured broadly in the direction of the window, "there, a mere one hundred yards from this spot, occured what will surely be recorded as one of the most unfortunate," the bar officer, a heavy smoker, had begun to cough, "events in modern history." He stopped momentarily to stub out the Galoise with his stained left index finger. "There, over there, where the bastards" – he stopped again, this time for a shot of port – "couldn't wait for Home Rule like civilized human beings, where they started a bloody war, the Eastern Rebellion or some such nonsense. 'We serve neither King nor country, but Ireland,' they claimed. Anti-treaty forces, my ass. A bunch of rag-tag, hooligans that's all they were. Sheer nonsense, the whole bit of it."

"Where is everyone?" Jack asked again.

The bar officer must have heard the question since he was standing three feet away, but once more he chose to ignore it. "And then, when the war starts, they won't raise a bloody hand to help us. This neutrality nonsense was de Valera's idea," he continued, reaching for another cigarette. "He keeps changing his mind, the blooming idiot. Typically Irish – doesn't know which end is up! One day he's IRA and the next he's banned the blokes! He's one of the chief reasons that we lost the south – do you know that chaps? A bloody traitor, but neither Chamberlain nor Churchill were wise to him."

"Who?"

"de Valera, man! So what does Chamberlain do? He gives away the bloody Irish ports, ports that would have saved the lives of more than a few of our chaps. And then what does de Valera the bastard do?"

There was no reply.

"He goes and joins the bloody Germans, that's what he does."

"I thought Ireland was neutral," Jack remarked timidly.

"Rubbish," Covington protested. He finished the rest of his drink and poured another. "They claim they're neutral of course," Covington was beginning to squint again. "They claim they'd enter the war on our side if we returned the north, but that's a lie. They're pulling for the Jerries to win it all – that's what it's really about! You wait until you see all the bloody guards chumming up to Jerry next door."

"Why are they so friendly with the Germans," Keefer asked.

"So they can be free of the poisonous yoke of British imperialism. My goodness, you chaps have a lot to learn."

"I thought they were already independent," said Keefer.

"Yes, and no."

Jack looked sceptically at the British officer. "Sounds a bit odd, Covington, I mean how do the Irish people feel about all of this? You make it sound as if this guy de Valera is ..."

"You're bloody right he's crazy. He's IRA, isn't he? And I'll tell you something else. Most of these people," he gestured around the bar as if the bar was full of Irishmen, "support us and hate the Germans. Why else would thousands be going to England to enlist?"[8]

After a moment or two, the Canadians nodded agreeably. Even Jack, with the critical eye of a reporter, was prepared to assume that Ireland was just like Canada – a loyal dominion of Britain – until it

8 While the rest of Covington's historical analysis might be open to question, this point isn't. Over 50,000 Irish crossed the Irish Sea during the war and enlisted on the Allied side.

had been proven otherwise. Any sane person knew the Germans were the enemy, not the English.

It took the three internees a while to finish their first bottle of port before moving on to the second. The Canadians were drinking slowly, though picking up speed as the afternoon wore on, while the bar officer maintained a steady pace, though he seemed less and less enthusiastic about dipping into the inner cabinet – and more and more in need of convincing that the occasion was still auspicious.

After a while Keefer looked at his watch; it was already 3:30 in the afternoon. He got up and stretched, poking his head outside. It was raining and the camp was still quiet, with little sign of life. When he came back, Covington had poured the remains of the second bottle into Jack's glass, and was reaching behind him for a third.

"Do we ever get a look at these German bastards?" Jack asked, supporting himself on his elbows and tapping his glass on the bar.

"Jerry is over that wall," Covington replied, pointing out the window. "We rarely see them though. Occasionally they hang over to cheer whenever someone is captured, or we throw rocks at each other, but otherwise they keep to themselves. The bastards like it here," Covington grinned.

"Bloody cowards," added Jack affably.

Covington nodded half-heartedly at Jack. "But no fraternizing please, gentlemen, particularly if you bump heads at a public house."

"Public house?"

"Yes, in one of the towns nearby, Newbridge,[9] or Naas, that sort of thing."

"You mean we get to leave the camp?"

Covington stopped in his tracks, realizing that he had inadvertently raised the topic that he was intentionally avoiding. "Yes, whenever you want."

"*What did you say*?" asked Jack, the bar officer's words having finally sunk in.

"When we sign our parole, Jack, we are under strict orders to ignore the Germans."

"*Parole*?"

9 More often known, today, by its Gaelic name, Droichead Nua.

"Yes, when you are on parole – avoid the Germans," Covington repeated simply. "And if you haven't figured it out yet, that's where everyone is."

The Canadians sat silent and confused, the incongruous notion of freedom in such a setting beyond their grasp. Waking up amidst the barbed wire and armed guards, only to head out for a tour of the Irish countryside an hour later, topped off by an evening in the local pub, cozying up to your favourite oberlieutenent? One minute bombing the hell out of each other, the next fighting over last call? What was this about?

"Every so often we bust out, dig a tunnel, or throw ourselves into the barbed wire, but it's tough, I tell you," said Covington. "We've tried bribing the guards, tying them up, stupefying them, hiding our parole oaths, hiring secret agents, approaching gentry friends for assistance, but nothing works. It's frustrating as hell."

"Okay, but ..."

"And I tell you, Jack,“ Covington said, continuing to steer the topic away from where it was inevitably headed, "having Jerry next door hardly helps matters. Earlier this year when the *Bismarck* sank the *Hood*, those idiots spent the whole bloody night singing, and carrying on – "Deutschland uber Alles" till dawn! – shouting at us from the tops of their roofs like a bunch of Americans. If only we'd known that you brave Canadians were up in the sky you could have ended the celebration for us, eh wot ... But then a few nights later we returned the favour when our boys sank the *Bismarck*. Since then we've been collecting small rocks near the fence and one of these days we'll let them have it."

Jack was fighting frustration and a rising temper. It seemed as if every permanent officer he'd met was crazy. "Aubrey ..."

"Covie, please Jack."

"alright, Covie. Tell me this much: is parole pukka gen or duff gen?"[10]

"Sounds pukka to me," said Keefer.

Covington reached for another Galoise. "Chaps, I was going to leave the matter of parole to tomorrow, with today as a get-acquainted sort of day."

"No Covie, you were going to tell us about parole today." Jack

10 RAF expressions: pukka gen was good news, duff gen was not.

drained his glass, slamming it down on the bar.

"Yes, but it's better to break you chaps in slowly."

"Why?"

"Because it just is. Parole is a difficult concept to accept, particularly for you chaps, being *volunteers*, and all."

In fact Covington, as the senior officer in camp, had been warned the previous night. He wasn't the only one worried that the Canadians might reject the notion of parole, bringing disgrace upon themselves and the others. Covington would have vigorously disputed that this worry was motivated by self-interest, for in truth he could see no good coming to either group if they failed to tow the line by simply leaving camp and heading for Belfast.

"Very well, then." He put down his drink and spoke slowly. "Parole is a conditional release from camp. It is a privilege granted by the Irish authorities to us and the Nazis alike. Typically, we sign a slip at the parole hut which will permit us to be at large until 02:00 hours each morning, seven days a week. Our oath requires us to return after that. It's been agreed to so there is no sense arguing."

"You mean to tell us," Jack said, "that tomorrow morning we can go over there, sign a piece of paper, and walk out of camp."

"Yes, Jack, You can go today if you want."

"Bobby, my man," Jack said, raising his glass, "Belfast, here we come."

"It's not that simple, Jack," snapped the bar officer. Calder stared back at him. "It's not that simple," he repeated. "When accepting parole, you give your word as an officer and gentleman of the Royal Air Force that you will honour the terms of your conditional release." Now he was really squinting. "You must return to camp. Otherwise, you will bring dishonour not only to yourselves and your crew but also your country."

"There's a war on, Covie," Jack reminded the Englishman, "and we haven't come 7,000 miles to play politics."

"Yes, Jack, but you have another problem." Covington paused to straighten his tie and the collar on his tweed jacket, looking nervously towards the port. "You see, Canada, through High Commissioner Kearney, has agreed to honour the arrangement, so you're stuck here."

"Yeah, but what about the RAF?"

"What about them, Jack?"

"Wait a minute," Keefer said. "You mean the RAF has agreed to this."

"In a way, yes."

"The Jerries are marching willy-nilly through Europe," Jack had stood up from his bar stool, "and you're telling us that we're stuck here for the rest of the war."

"Maybe ... but you can escape with the approval of the RAF as long as you don't do it on parole."

"What?"

"For instance," the English officer continued, determined to get his point across, "once your parole is cancelled and you have returned to camp, you can escape. In fact, they want us to escape."

"Doesn't sound like it to me," said Keefer.

"You see, the barbed wire is out there for a reason and that is to remind us of our status here and to provide a proper obstacle for an escape." He was now shaking his head from side to side and rubbing his eyes. "While outside, you shall not do anything to plan an escape – you can't even talk about it – and you cannot ..."

"How are they going to enforce that?"

"That's not the point, Jack. The point is that we're not entitled to enlist local assistance from the Protestant gentry, although most of them support us."

"You're crazy, you know that."

"Not my rules, Jack."

"Enlisting local assistance,“ Keefer snapped. "You just said that everyone supports us, so why the hell ..."

"Well, to be honest, chaps, we break the rule against enlisting local assistance every so often, but the RAF ..."

"What's the worst that could happen to us if we escaped on parole," Jack asked.

"Don't ask," Covington replied.

"I don't know much about this place, Covie 'ole bean," Jack said, "but I know that it's our duty to escape no matter where we're imprisoned. That's RAF code man – our sworn and sacred duty. No one can punish us for performing our goddamn duty."

Covington stared unhappily at Calder for a few seconds, and then at Keefer. He was displeased by their insolence, by the irreverence he saw from all the volunteers, but from these two in particular. He couldn't understand why the RAF would appoint two people like them to manage a bombing crew, nothing but farm boys from Canada who thought they were so grand, so important to the war effort, boys who probably didn't know where England was before 1940, who probably grew up on a North American Indian reservation with naked savages, ignoramuses who couldn't find Ireland on a map even if their lives depended on it (but who had somehow gotten lucky this time). He knew that their appointment was part of a political compromise to ensure that the RAF would obtain a steady supply of pilots since so many English ones had died. But couldn't they do any better than these idiots?

"If you escape on parole," he said flatly, "the RAF will send you back to the Curragh. Thereafter, you will be marked as insubordinate trouble-makers. Now have I made myself clear?" He stubbed out his cigarette and faced Jack directly.

"They should give us a medal for chrissake," Jack said, defying him.

"They'll court-martial you," Covington shot back. "Do you understand that Calder? L.M.F."[11]

There was a brief pause as the bar officer reached down, opened the cupboard of the inner cabinet, and placed the remaining port away on the top shelf. Sir John Maffey, the British representative to Eire, had brought it up on his weekly visit. A couple of Canadians, one a well-known reporter, the other a former football star. God help us.

"That's the biggest bunch of bullshit I ever heard," said Jack. In a matter of hours he had gone from thinking "who do these Irish bastards think they are," to "only the English could have thought this up."

He stood up and faced Covington, fists cocked, ready for battle.

"Okay, okay, let's calm down now," Keefer said, grabbing his navigator by the collar. "Come on, Covie, if we can't 'enlist local assistance' as you put it, then how in hell are we going to get to the North?"

"Well, Keefer, there are ways," replied Covington slowly, watching Jack and keeping his distance. "For instance, if you strike up a friendship with an Irish girl, and she offers to help you, you are

11 Another Air Force expression: lacking in moral fibre, normally reserved for bed wetters.

obliged to politely decline, unless you marry her – although she's not likely to unless she's Catholic, so you'll have to convert – but in any event if you do convert and marry her and live with her on parole, then you'll have a good chance to make it to the North because her parents ..."

"What the fuck are you talking about?" Jack screamed at the top of his lungs. "I don't want to get married – I just want to go back to the war!"

"Back off, Calder," Covington shouted back, "I don't make the rules."

Things had definitely gone a bit crazy. Covington and Calder were now standing up, pushing and pulling at each other's lapels, with Covington muttering that he hated to see things get off on the wrong foot like this but that it didn't surprise him either, while Keefer was doing his best to keep things under control. After Keefer had pushed them apart, Covington turned and pleaded directly with him. "What I'm trying to tell your thick friend is that there is only one way to get out and stay out of the Curragh, and that's by busting out. If you haven't signed a slip, or if they've cancelled it at the end of the day, then they can't send you back. But life isn't so bad here, Keefer – do tell the idiot that. You'll meet lots of girls on parole, the gentry is here in full force, the riding is excellent, there are horse meetings and point-to-points every weekend, excellent fishing down by the Liffey, golf, that sort of thing. We stay as busy as we can. I've tried to escape a few times; believe me, it's not easy."

"Then why don't you walk out of the fucking gate?" snapped Jack, moving once more towards the bar officer.

"Jack, sit down," said Keefer.

"You still don't get it," Covington replied, placing his glass down on the bar. The Englishman was now preparing to battle to defend the name and rank of all permanent RAF officers everywhere. "I've heard about you Calder, you and your loose lips – as if someone of your ilk could ever write anything of even marginal intelligence. Call yourself informed, damned if I read any more of those silly Canadian papers!" Sir John had provided them with a copy of the *Gneisenau* story, thinking it might be a worthy gesture to welcome the new arrivals. Little did he know how the English officers would react – "'RAT-TAT-TAT. I got one. I got one. Slapsy Maxie. Soooo cool, sooo cool! Oh, yes! Oh, yes! A regular line shooter.' Short silly little sentences, whatever happened to a proper subject and a predicate? A decent clause? Bloody North Americans ruining the language, don't they have schools over there? Something you'd expect of the bloody Irish! A famous Canadian reporter? God help us!" Covington raised his fists and began circling his target.

"OK, OK, Covie ..." urged Keefer, shaking his head.

"And I can assure you that if you write anything about this internment business, you'll be thrown out of the RAF and sent home in disgrace. People don't need to read about us." Covington then stopped, facing Jack directly, crouched in a drunken, fighter's stance, swaying unsteadily back and forth on his feet. Suddenly he lunged at the Canadian, unleashing a wild left cross at Calder's head with his Galoise leading the way. But Calder stepped back quickly, feinted with his left, and hit the bar officer with a right, just above Covington's ear. Covington fell back against the bar, knocking over the three empty glasses which then shattered on the floor.

Keefer bent down to pick up the broken glass. He then pushed the Englishman away and ordered his navigator to sit down. "We were talking about parole, Covington, not newspapers. All we want is to get the hell out of here. You can understand that, can't you."

The bar officer was breathing deeply, running his hands through his hair and rubbing his face. He remained that way for a full sixty seconds before offering the final word: "There is only one way to escape, gents – through the barbed wire. There's nothing you can do about it and it's frustrating as hell!"

4 LIFE ON THE CURRAGH

Heading back to his hut, Jack Calder was feeling a little guilty. It wasn't only the botched mission to Frankfurt and his dubious role in it – or even his scrap with Covington, the camp's senior internee and conscience, as he would later discover – that bothered him. Admittedly, one hated to get off on the wrong foot with these guys, the permanent types, particularly those senior to a lowly pilot officer cum washed-out navigator, the lowest of the low when it came to the ranks of the privileged. Nor was Jack quick to pick a fight. In fact he rarely fought, even if he considered himself part Irish (with his long lost aunt and all), even if he considered himself as rough and tough as the next guy, buying the simple notion for the time being of Paddy lovin' a good scrap above all else. He just lost it, that's all. And who could blame him? It had to be the craziest camp he, or anyone else for that matter, had ever encountered.

No, it was something else, he realized, another reason to feel that way. For, as a young man, Jack Calder had always wanted to be a reporter, through his graduation from the University of Toronto with honours to his first job with the Chatham *Daily News*, covering high school hockey games, town meetings and sidewalk closures, on up to his lofty editorial position with Canadian Press, Canada's national wire service. Yet, what was he to make of all of this? Irish neutrality wasn't exactly hot news back home. The isolationists of the world weren't exactly grabbing centre stage. Who would be interested in the details of this bizarre, sorry place?

The man was clearly in a funk.

Part of the problem, it should be explained, was that no one in his profession, Jack included, saw his or her job in modern terms back then. In another era, what had just happened to him and Bobby would have been newsworthy. In the middle of a war, however, all the news that's fit to print was not what people needed. What most of the allied world preferred to hear and read about instead were acts of heroism: the battles, the predictions, the victories, in all their glorious detail. The stories of brave men and women rising above their circumstances, their sacrifices. That's what made the news. And you only found that on the front lines: at sea, on the ground, or in the air, through the eyes and ears of the war correspondents. The heroes of the profession. Men like William Morrow and Bill Shire, or John Steinbeck and Ernest Hemingway, who would go on to even greater careers as writers later in life. That's where Jack's path led. He had, after all, become the first journalist in North America to cover a bombing mission from where it counted most – looking down the chute. He'd love to be back up there at 20,000 feet, supporting the war effort as best he could. Heady stuff for a young guy who, as a copy editor at CP a year earlier, had virtually trained himself: correcting the mistakes of his fellow reporters, drafting the headlines, determining what was newsworthy and what was not. And much was noteworthy. Not just the big riot on Ste-Catherine Street when he'd met Keefer a few years earlier but all the monumental changes after that: Hitler's annexation of Austria and Sudetenland when peace with honor was still an option, Britain's declaration as the *Athenia* and the *Courageous* went down; Prime Minister King's leadership, the pride and outrage with which Canada entered the war, watching as each loyal Dominion lined up dutifully in defence of the Crown. What Covington, a brave fighter pilot in his own right, had just offered Jack from his usual post at the Officer's Bar – the Easter Rebellion, the raising of the tri-colour, the anti-treaty forces, de Valera's treachery, even parole – seemed so pale by comparison, so trite. Neutrality might have been a topic of interest elsewhere, but not in Canada in the fall of 1941. There was no longer a raging debate between the Lindbergs of the world, and the rest. That was history.

The bit about the Germans next door, now that was interesting. But what could be made of that, really.

The Vancouver Sun — EXTRA

Only Evening Newspaper Owned, Controlled and Operated by Vancouver People

FOUNDED 1886 VOL. LIII—No. 279 | OFFICIAL WEATHER FORECAST Cool; Scattered Showers | VANCOUVER, BRITISH COLUMBIA, SUNDAY, SEPTEMBER 3, 1939 | Price 5 Cents | Trinity 4

BRITAIN AT WAR!

Ottawa Rushes Preparations

Cabinet Meets As Australia Declares Wa

By British United Press

CANBERRA, Sept. 4 (2:30 a.m. Monday) — Aus tralia declared today that a state of war exists betwee her and Germany.

By Canadian Press

OTTAWA, Sept. 3.—Dominion cabinet min ters, aroused early by the dread but not unexpect news that Great Britain and the Empire are at w with Germany, hurried through foggy streets tod to meet Prime Minister Mackenzie King in the Pri Council chamber at 10 o'clock (6 a.m. Vancouv time).

Mr. King, advised by the Canadian Press a fe minutes after the flash was received that Gre Britain had declared war, lost no time in commun cating with his ministers, who had been warned t be ready for such an emergency.

Hon. Ian Mackenzie, minister of defense, was the fir to arrive for the council. He reached the meeting at 9:45 a.m. Shortly afterwards Hon. C. D. Howe, minister of transpor Hon. Ernest Lapointe, minister of justice; Hon. J. G. Ga diner, minister of agriculture; Hon. Norman McLarty, pos master-general, and Hon. T. A. Crerar, minister of resource arrived.

Chamberlain Tells Empire of Decisio

France Joins in Declaration Against Germa

Prime Minister Predicts 'A Liberated Europe and Hitlerism Destroyed'

By WEBB MILLER

LONDON, Sept. 3.–Great Britain went to war ag Germany today–25 years and 30 days from the time she tered the conflict of 1914 against the same enemy.

A brief announcement by Prime Minister Neville Ch berlain that went by radio to all outposts of the Empire Britain to war in fulfillment of her pledge to help Polar that nation was invaded by Adolf Hitler's Nazis.

The French government set its deadline at 5 p.m a.m. Vancouver time) but announced from Paris that Fr considered herself automatically at war with Germany moment Chamberlain made his pronouncement.

"This country is at war with Germany," Chamberlain said in measured tones. "You can imagine what a bitter blow this is t that all my long struggle to win peace has failed."

A radio hook-up to all places under the Union Jack was mad Chamberlain stepped to the microphone in No. 10 Downing Str speak the fateful words.

"We have a clear conscience," declared the Prime Minister. "W done all that any country could do to establish peace, but the situati become intolerable, and we have resolved to finish it.

"Now may God bless you all and may He defend the right, fo evil things that we shall be fighting against — force, bad faith, inj oppression and persecution. Against them, I am certain, the right will pr

"God Save the King" was played on the BBC's Empire hook- Chamberlain concluded.

Wounded Polish Airman Battles 12 Nazi Planes

WARSAW, Sept. 3 — Lieutenant Pausinski of the Polish Air Force was hailed in Warsaw today as an early hero of the war.

Throng in Downing Street

EDITORIAL

Fighting For a Just Cause

You are to be proud today of your citizenship in Canada and the British Empire.

Was there ever a war so just as this one to which we are solemnly committed?

Has there ever been, in the world's history, a more noble event than for Britain and France, as they decide today, to come promptly to the relief of their ally, now in dire need of succor?

Selfish interest might have called for our two allied peoples to save their own skins, even at the sacrifice of their treaty undertakings. It is something akin to this sentiment that is relied upon by the isolationist peoples of the world today, as they seek to justify a position of aloofness.

We belong to an Empire and we belong to a breed which honors its commitments. Poland's fight, and what it stands for, is our fight today.

In Germany, the Allied Front, for what it means, will not be completely understood. The Sun has already noted that the German mistake of 1914 is being repeated 25 years later. And for the second time the lesson must be brought home to the German people. Another generation has been led away—this time by Hitler and a gang of cold-blooded adventurers who already have committed every crime of rapacity and oppression that is listed in the calendar.

In London, this noon, Mr. Chamberlain stood in his place in Britain's Parliament, as chosen head of the free people of this Empire, announced that we are at war with Germany. His was a tragic and difficult task; and we shall all of us face difficult tasks before this conflict is ended. But today we have a great satisfaction. It is that with infinite restraint and patience, we in this Empire, through our chosen leaders, have sought to intervene by every peaceful means that could be summoned to our aid. We have tried to appease and placate and advise; in every way, over a long period, we have thrown our weight and influence on the side where right is not the sole prerogative of might. We have given "last warnings" and have delayed more precious hours to allow those warnings to sink in with full effect. But Herr Hitler, holding to his record of duplicity and grab has chosen to go unheeding the other way.

Thus, we have arrived, sadly but still firm in resolve, at today's fateful decision. We shall have no fear of the outcome. That is not the British way. It is a part of the propaganda of the Hitler-Stalin ideal that the democratic way of living is to be swept aside for that nameless shambles of ruthlessness and disorder which has reduced the peoples of Germany and Russia to practical serfdom. Don't be fooled by this nonsense! We are facing days of personal worry and national trial, but the calmness and common-sense of mankind will ultimately prevail. There shall be no other end.

Again we say we have a just cause and a clear national conscience. Last Sunday, the clergy of Vancouver and other cities in our land led the people in prayers that we might be delivered from the horror and suffering of another war. Today, we shall devoutly pray again for guidance and for victory over the evil forces which stalk through the world, seeking to ruin the liberty and decent way of life of mankind.

Lindbergh Rall Scene Of Clash

NEW YORK.—(UP)—Seve persons were injured Wedn day night in front of Manhat Centre while an "America Fir rally was in progress, address by Col. Charles A. Lindbergh

Pickets representing the "S dent Defenders of Democrac attempted to pass out handbi opposing the rally to a str crowd. Some persons in t crowd attacked the picketers.

Approximately 10,000 pe sons, who were unable to g into the rally advocating Amer can neutrality, gathered in th street in front of the buildin to listen to the addresses ove loudspeakers.

Denounce Speakers

The pickets, carrying placard denouncing Lindbergh and th other speakers, began a demon stration but were forced away by police.

Col. Lindbergh told the "America First" rally it is now obvious that England is losing the war" but has "one last desperate plan" to persuade the United States to send another American Expeditionary Force to Europe.

He accused Britain of misinforming the United States and other nations "concerning her state of preparation, her military strength, and the progress of the war."

Ireland Wants To Be Neutral

By Associated Press

DUBLIN, Sept. 2.—Prime Minister Eamon De Valera told an emergency session of the Dail (Parliament) today Eire would try to maintain neutrality as long as possible in any European conflict.

Mr. De Valera summoned the Dail to pass emergency laws for control of food, currency, transport and shipping.

Vancouver Sun, 3 September 1939. The pride of citizenship in the Empire described so eloquently in the *Sun*'s editorial above was sometimes tough for his audience to grasp – though the evil of Hitler never was.

So therein lay the problem. A reporter's job was to report, yes, but not if neither the reporter nor his audience was interested. It didn't leave many options for a guy with a restless mind, down on his luck.

As for Keefer, he stumbled back to their hut that night dodging wire, feinting imaginary guards, and dropping his shoulder hard into his bed. After a fitful sleep, he awoke disoriented for the third consecutive morning. He didn't know why, but he suspected later that it might have been the result of the first of several recurring dreams he had thoughout his internment – filled with guilt, self-recrimination, and an intense desire to go out and get himself killed. It was enough to make anyone hungry.

He immediately set out for some food, but landed up in the NCO's mess instead.

"Good morning sir," said a surprised Dalton, as Keefer appeared at the table closest to the door. Diaper and Brown smiled uncertainly and Virtue continued eating. "Not that we mind at all sir, but the officers' mess is on the far side of the hall, just as you enter, sir."

"I'm aware of that," replied Keefer, feeling as if he had just picked up a fumble and run it into his own end zone. Still, Virtue's breakfast looked good by any standards: bread, bacon, eggs, toast, an assortment of jams and marmalades, and tea. He pulled up a chair. "So how are you feeling, Virtue?" he asked. His tail gunner still had his head wrapped in a head bandage, courtesy of the NCO's rifle butt in Limerick.

"Not bad, thanks." The tail gunner stopped eating. "You know, sir, it was awfully good of you to come back and free me as you did. You saved my life."

"Not at all, Virtue, not at all." He looked over at the other three sergeants, Diaper, Dalton, and Brown, who were also nodding respectfully in his direction.

"It wasn't your fault we overshot, sir," added Dalton. "It could have happened to anyone."

"Yes, thank you, Dolly."

"Have you been interrogated yet sir?" Dalton asked, changing the subject. Keefer wasn't sure if his wireless operator was referring to the previous day's pub crawl or not. "Rumour has it that the senior officer of each crew is interrogated by this chap named McNally, the camp commandant. Yes sir. Some of these gents here," he motioned to the twenty-odd NCOs present in the mess, "say that this McNally can be pretty tough." Keefer looked around the room, first at the NCOs staring at him and then at his crew, who were nodding in agreement. "Apparently he is quite different from the LDF," continued Dalton. "He's a teetotaller, very favourable to the Germans."

"That's right, sir," Diaper, his second pilot, added. "They say he's an odd duck."

"Odd duck?"

"Yes sir," Diaper replied. "Not above pressing his point."

"Thanks lads, I appreciate the tip," Keefer said, rising from the table.

"Oh, and sir?"

"Yes, Dolly?"

"Best of luck finding your way back to the hut."

After a similar breakfast in the officers' mess, Keefer was greeted by an Irish corporal who announced that Colonel McNally, the Curragh's commandant, had requested the pleasure of his company – something he was now expecting. The corporal asked if he might be permitted to show Keefer the way.

"May I have fifteen minutes to wash up?" Keefer asked meekly.

Keefer burst into their hut. "Get up, Jack. My interrogation starts in fifteen minutes."

Jack lay motionless on his bed. Maybe he drank himself to death, thought Keefer. "He's going to interrogate me, Jack. He'll want to know why we were in Irish airspace. What should I tell him? What should I tell him?"

"Don't tell him you're from Boston," Jack mumbled, alive, after all.

"Jack, I'm serious." Keefer went to shake him. "They might torture me. What do I tell him?"

"You're crazy – now let me sleep."

"Seriously," he pleaded.

"Tell him anything. Tell him you'd had enough of the war and you wanted to meet some Irish girls – just leave me alone."

He would tell him nothing, Keefer decided, as he walked out in disgust, through the compound and out to the gate where he met the corporal, who ushered him through the outer compound, past the German camp, to the commandant's quarters. He'd let this guy McNally suffer, just as he would suffer. He was even prepared to be tortured, if that's what it took to survive with the honour and dignity of an RAF officer intact – which was more than he could say for his navigator, who wouldn't even get off his goddamn bed to help him.

"That's the problem with you journalists, Jack," he snapped contemptuously, slamming the door to their hut behind him. "You're moody, and you drink too much."

A moment later he strode up to the oak door of the commandant's hut, and pounded confidently. He heard footsteps approaching. A kindly looking man about sixty years old appeared. He had thin grey hair combed neatly down on his scalp and was dressed smartly in a grey wool v-neck sweater, brown wool tie, beige military shirt, khaki green trousers, and brogues. He wore horn-rimmed glasses and stood only five foot eight. "Good morning," he said, introducing himself as Colonel Thomas McNally and offering Keefer a cup of tea. The interrogation had begun.

Keefer declined the tea and remained standing by the door.

The colonel shrugged, and sat down at his desk. "Pilot Officer," he paused to look down at a paper in front of him, "Keefer, then is it?"

"Yes sir," the pilot officer replied, smartly.

Colonel McNally nodded. "Relax lad, come over here and have a seat."

"No thank you, sir."

McNally shrugged once again. "Tell me Keefer, are you feeling settled in your new home?"

"Sir, my name is Pilot Officer Bobby Keefer."

"I just said that, for the love of Jesus. Now, are your accomodations satisfactory?"

"I'm stationed with the 103 Squadron, Royal Air Force, Elsham Wolds, England, sir! My date of birth is 28 September 1917. My identification number is J.4.8.7 ..." The pilot officer, off to an impressive start, began fumbling nervously for the dog tag around his neck.

"What are you doing?"

"I am abiding by the rules of war, sir, and I expect you to do the same." Keefer had been waiting for an opportunity to say that. His confidence soared.

"What in God's name are you talking about?" the colonel asked, as if his feelings had been hurt. "And you don't need to shout! Come over here please."

The pilot officer remained standing but shuffled a few feet over towards the desk.

"That's better." Just my luck, thought the colonel, a Canadian. "I wanted to tell you that I'm at your disposal should you require any thing at all, any thing at all. If you have any problems with the guards, I'd like to know about them."

"My identification number, sir, is ..."

"Come, come lad, you can't be serious," McNally repeated impatiently. "Then at least listen. I know you'll be frustrated here, but we'll do the best we can to make you feel at home. Do you have any questions?"

Keefer couldn't be fooled that easily. He watched as the colonel stared out the window.

"Och, I see Covington starting out for the Liffey to drop a line."

Keefer turned his head and spotted the bar officer on his way out to the main gate.

"Do you ever drop a line, then?" asked the Colonel absently.

Keefer turned back to his interrogator, eyes fixed firmly ahead.

"The cat got your tongue? Fine. I want to warn you about parole. You're to return at the designated time, that's all there is to it. To tell you the truth, I couldn't give a tinker's damn what time you get back but if you're late, the diplomats and the Dail will have something to say about it. Understood lad?"

There was no response.

"So! You're just going to sit there? Fine. We've got Germans too if you haven't already figured that out, and they have to be back in a timely fashion as well. You'll be treated the same in all respects. Behave yourselves when you're on parole. The RAF doesn't want you socializing with the Germans and the Luftwaffe doesn't want them socializing with you. But that's up to you. I couldn't care less – no fights, that's all. I don't mind you trying to escape, that's your duty, but no beating on the guards. And tell your men to watch their language. And loosen up a bit, for the love of Mary!"

Keefer, thinking he may have been had, was beginning to feel stupid. "Yes sir, thank you, sir." He was now standing in front of the desk. "To be honest sir," he tried to look as proud and defiant as possible, "if we're going to be kept here as prisoners, then I feel we should play the part. In short, I won't be taking parole, colonel, nor will any of my men."

Colonel McNally smiled, glancing out the window at Covington, receding into the distance, pole in hand. "We'll see how long that lasts." Having served the Irish army for twenty-five years, the colonel had seen it all.

"Is there anything else, sir?" Keefer asked awkwardly.

"No, I guess not," the older man replied. "I mean what I said, Keefer. Feel free to come by and have a chat, and tell your navigator too. I understand he used to be a reporter. Please no stories. Anyway, I do hope you enjoy it here. I understand how you Canadians must feel, coming all this way to fight."

He sighed and stood up. "That's it, then. I wish I could go fishing in the middle of a working day like you, but I have things to do." And with that Colonel McNally led the pilot officer to the door, concluding the interrogation.

"You told him what?" asked Jack a moment later, after Keefer had returned to their hut.

"I told him that we were prisoners, and that we expected to be treated like prisoners. So no parole. And no articles. Got it, Jack?"

"Fine." Keefer might as well have waived a red flag in front of a bull. "But no parole? That's pretty extreme."

"Like you said, Jack, we didn't come 7,000 miles to play politics. If the Brits think it's all right to go out picnicking or hunting foxes, or whatever it is they do, that's up to them. But that's not right for us."

"Yeah, but no parole ..."

"Parole is bullshit. It's a bribe, that's all it is. If we take it everyday, sooner or later we won't feel like escaping."

Jack was giving the matter some thought. "Then why don't we just go to Northern Ireland, just leave?"

"You heard Covington. We'd be discharged LMF for sure."

"Then let's go back to Canada. You know how long some of these guys have been here? What are our choices? Staying here and throwing ourselves into the barbed wire every night? Let's go home."

"We can't do that, Jack. No one would speak to us again."

The two looked off in separate directions, knowing that what Keefer said was true.

Although its exact duration became a matter of some doubt to his audience over the years – stretching from days into weeks, then from weeks into months – Keefer's protest, or conscientious objection if you will, lasted exactly seven days.[12] The problem was that his life had suddenly slowed down. No more hair-raising all-night bombing runs to look forward to, watching in fascination as the tracers skipped across his path. No more thrill of the drop, thinking that he might single-handedly win the war (while it was a gruesome business, it was damned exciting!). No more storming Keith's brewery on Purdie's Warf with Jack, or toasts at the officer's mess to their fallen comrades, laughing at Lord Haw Haw's predictions of their demise. No more trips to London, to Minsky's or the Trocadero, where they'd wait backstage in the hope of meeting some chorus girls, hoping to find a go'er, (like the Regent, where one night they finally found a couple, only to be tossed out by the general manager before they could get into any real trouble). In short, no more freedom to do what they wanted to do, to do their duty. And he didn't like it one bit, for flyers are by nature a restive lot.

So eventually he and Jack found themselves together at the police hut, preparing for their first sojourn into the countryside.

"Have you seen one of these, lads?" asked Captain Frank Fitzpatrick in a thick Monahan accent early one morning the following week, blank parole form in hand. Corporal Fitzpatrick, in his mid-forties, was their parole officer, second in command to Colonel McNally. As a former internee of the English in Wales in 1916, Fitzpatrick was ideal for the job – he knew all the tricks.[13]

The captain was standing behind a customs-style counter in a small wooden hut, waving a black and white 5″ x 7″ preprinted document under Keefer's nose.

"This is known as a parole slip," the Captain continued. "Each of you is to sign on the dotted line." He then pointed to a signature line somewhere in the middle of the document, with written instructions above and below. "But only after I write in the time of your departure," which he did, "and the time of your return which will be, what, lads, 02:00 hours?" They nodded. "When you get back to camp, I'll take your slip out of the drawer and stamp it" – he showed them a stamp marked 'cancelled' on the counter – "and then I'll keep it."

12 The first parole slip on file in Keefer's name at the Irish Military Archives at Parkgate, Dublin, is dated 2 November 1941.

13 Not that the Irish were ever granted parole by the English during the Easter Rebellion, something the guards frequently pointed out.

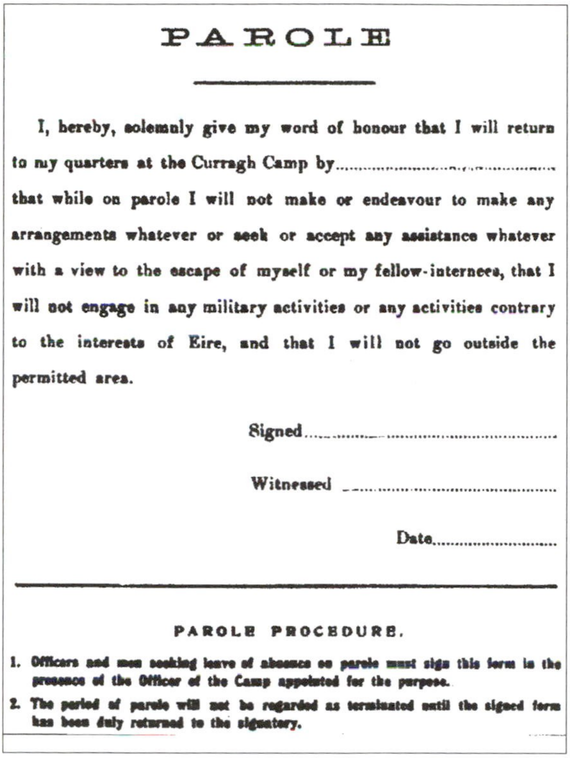

PAROLE

I, hereby, solemnly give my word of honour that I will return to my quarters at the Curragh Camp by.................................... that while on parole I will not make or endeavour to make any arrangements whatever or seek or accept any assistance whatever with a view to the escape of myself or my fellow-internees, that I will not engage in any military activities or any activities contrary to the interests of Eire, and that I will not go outside the permitted area.

Signed..

Witnessed ....

Date............................

PAROLE PROCEDURE.

1. Officers and men seeking leave of absence on parole must sign this form in the presence of the Officer of the Camp appointed for the purpose.
2. The period of parole will not be regarded as terminated until the signed form has been duly returned to the signatory.

The careful reader might wish to note a few things. The "permitted area" referred to in the oath portion of the form was never specified in writing and was left largely to the discretion of Colonel McNally. Also, the two paragraphs of fine print regarding "parole procedure" were changed as a result of events to follow. Courtesy Irish Military Archives

"And then what happens for the love of Mary?"

"Don't get lippy with me, Midgely, or I'll box your ears."

Joining the two Canadians for their first day of parole was Pilot Officer Dennis Midgely, another Englishman, one of ten fellow RAF officer internees as of November 1941. Midgely had kindly agreed to arrange a round of golf for the two Canadians.

"It's Pilot Officer Midgely to you, Captain."

"In your dreams."

Many of the RAF officers, especially the English ones, felt that their gaolers were obliged by military tradition to recognize their rank, regardless of their stature as internees. This view was not shared by the guards themselves.

"As I was saying before I was so rudely interrupted," continued Captain Fitzpatrick, sneering at the English officer, "after your slip has been returned and cancelled, you can throw yourself into the barbed wire all you like but not before. Got it?" Captain

Fitzpatrick, who had yet to meet Keefer, wanted to be sure that the pilot officer understood.

Keefer nodded respectfully.

"That's better. So why are you lads fighting for the English?" He had met Jack two days earlier, submitting to a brief interview. This time he'd ask the questions.

"At least we're fighting, you yellow bastard."

"Now, now, Midgely. That's no way to talk to a man with a rifle is it?"

"Drop it you Irish scum, and then we'll see."

"How can you expect me to sign your parole slip with language like that?"

Midgely, a scrapper from Yorkshire, had every reason to be bitter. While he had parachuted safely on the Salthills golf course in the west of Ireland, his second dickie, PO Edwards, hadn't been so lucky. He had drowned in the cold waters off Galway Bay, after the Irish hadn't answered their Mayday. Midge, as he was known, gritted his teeth and waited as Fitzpatrick slowly filled out the bottom portion of his slip, witnessing their oaths.

Pilot Officer David Midgely. Courtesy Bruce Girdlestone

"Acch, Keefer, I've heard about you. You're the Canadian pilot who was objecting or something like that." He dropped his pen and turned towards the pilot officer. "What's your problem? You don't like war? You don't like killing Germans?"

"I like war just fine, corporal," replied Keefer, anxious to be on his way.

"Then why are you objecting?"

"It's not war I'm objecting to, it's parole. Now could you please sign the slips and let us go."

"If you don't like parole, what are you doing here?"

"I reconsidered."

"Good for you, no point crying over spilt milk."

That Keefer's first day of parole would begin on the camp links was no surprise – golf being a lifelong passion of his. He had made the decision the night before. He and his navigator had played several times in the Wolds in central England that summer, at a course they had joined for a guinea a year, a special rate set by the membership committee in recognition of their volunteer service in the RAF, knowing that fewer and fewer airmen were living long enough to play a full season, whatever the dues. So golf seemed fitting, as if he was simply on leave or stood down, waiting for the weather.

The arrangement was that they would meet the club secretary, Major Whatcomb, who, thanks to Welply, would recommend Keefer and Calder for membership at one pound a month.

When the Major showed up at 11:00 hours, they proceeded to the first tee.

The camp links immediately next to the internment camp was remarkable, not only for its location but for its beauty. From the panoramic view from the first tee, where one could see large country estates buffered by gorse, heather and bracken to the cylindrical hollows in the fairways that marked where English army tents had been pitched during the Crimean War (making for ideal grass bunkers), the setting was addictive. When the frustration or boredom became too much to bear, the internees would often retire there, surrounded by only green and the sound of the wind whistling across the plain.

It was also remarkable, he soon discovered, for the fact that it was seldom used by others – the dream of every golfer. Given the fuel shortage in Ireland, no one could get there unless they were in the army, and soldiers were only allowed to play at restricted times. So the internees had the course pretty much to themselves, although they had to share it with the sheep. The internees and the sheep (which gave rise to predictable jokes after the war), hundreds of them on any given day, grazing through the fairways and rough, bleating like an army of rude spectators each time he addressed his ball, creatures totally incapable of understanding even the most basic etiquette of golf.

It took some getting used to, particularly when it came time to choosing a comfortable stance.

Clockwise: Keefer in uniform, enjoying some non-alcoholic refreshment. (They weren't allowed to wear their uniforms off the base except on the golf course.) Tents used during the Crimean war, courtesy Bill Gibson. The camp links (since renamed the Royal Curragh Golf Club, the name it bore prior to Partition) including clockwise, cylindrical holes in the 1st fairway where the tents were, present club house (the internment camp, which has since disappeared, was just beyond the club house), clock and water towers in distance, and commissariat butcher, now an abbatoir.

"That's all right, Keefer, simply move the ball," said the Major on the second hole after Keefer's long drive had faded right. "Winter rules, and all that."

He kicked his ball onto a dry, clean patch of fairway, grateful that he wouldn't have to spray the club secretary with sheep shit the next time he took a divot.

"But, what do you do in the summer?" he asked.

"Same thing, old boy – summer rules."

And they kept getting in the way. You couldn't exactly ask the sheep to let you play through. They wouldn't budge, even when you yelled at them, even when they were about to die.

"Jolly good shot," the Major exclaimed excitedly as Jack's drive on the ninth deflected with a dull crack off the head of a sheep, landing miraculously in the middle of the fairway, his best drive of the day. The animal Jack had struck, which had been grazing at the edge of the gorse, buckled at the knees.

"My God, what have I done?" Jack exclaimed, as the others ran up to check on the animal.

"You've killed it, my boy, that's what you've done," replied the major. "You're in, lads!"

"But what about the sheep, Major?" Jack asked. The major replied by ordering the three internees to drag the sheep over to the feed shelter by the eleventh tee. He then produced a can of red paint from the top shelf, together with an old brush, painted swastikas on the animal's forehead and backside, and propped it up by the door.

"Uhh … what are you doing, Major?" Jack asked.

"We'll just leave it here," he replied. "That way the farmer will think the Germans killed it."[14]

14 Whether his audience was being had on this one was never clear. There was one report in the archives that mentioned sheep and Germans in the same paragraph, but the rest of the report was indecipherable.

After lunch, courtesy of the major, the three internees returned to camp. Following some refreshment, duly noted in the bar officer's ledger, Keefer and Calder signed out two bicycles from a storage shed at the suggestion of another internee whom they'd just met over the past week, Pilot Officer David Welply, a cycling and tennis enthusiast. Welply was also English, and also a permanent member of the RAF.

This time when they passed through the parole or police hut Captain Fitzpatrick was nowhere to be seen. So, at Midgely's suggestion, they grabbed a form each, completed it, and then presented it to the sentry at the main gate. The system was, after all, based on honour.

Their intention was to visit a bank in Newbridge and the local tailor. In Keefer's case, having spent his last few pounds on membership dues and an old set of clubs (also courtesy of the major), he had been told that he could have his RAF pay wired there, care of the Northern Bank in Dublin. He figured that four pounds ought to be enough for a pair of tweed trousers and a blazer. While he had been comfortable in his flying jacket to date, he wasn't about to start turning his uniform inside out as some of them had for more formal occasions. He felt silly enough as it was.

And Jack thought he might look for a used typewriter.

"What for Jack? I thought you'd given that up."

"I just thought I'd write a few letters home. Okay?"

"Okay." Not that Keefer particularly cared one way or another, as long as his name didn't appear anywhere.[15]

After leaving camp the three turned south and followed a short straight road flanked on both sides by beech and oak trees. They soon came to the main base's water tower, passing the red brick and cedar barracks that housed the army proper, laid out neatly in rows like benches facing an altar, plus two identical churches, Protestant on the right, Catholic on the left, built in a time when there were still two religions in the south. The road then descended a gentle slope emptying out onto a plain, also known as the Curragh – from the gaelic, Curreagh Life, which meant a marshy place by the river Liffey. Part of the river in ancient times, the Curragh plain had

15 At that point neither Keefer nor the others knew about the *Maclean's* article, or how Jack had managed to smuggle the story out past the censors.

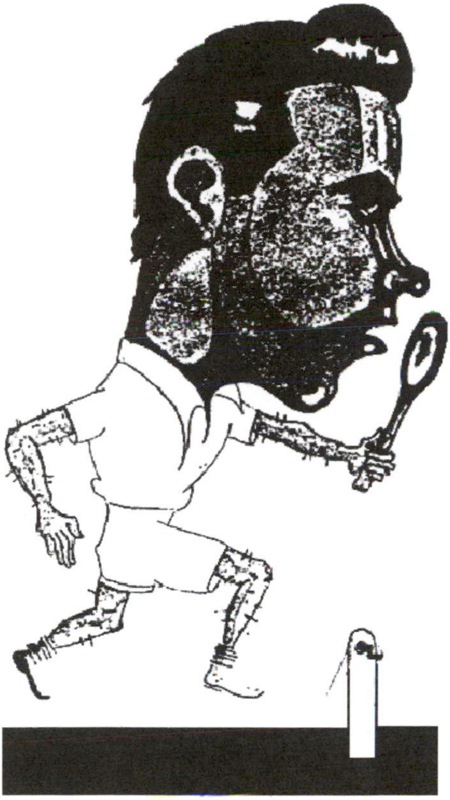

Top and left: Pilot Officer David Welply. (Courtesy Bruce Girdlestone.) After a smooth forced landing at Gark in county Sligo that January, Welply's aircraft, a Lougheed Hudson similar to the one pictured below, was given to de Valera to be used for the Irish leader's weekly flights to Berlin, or so the rumour went. Described in the *Irish Times* as "a great prize for the Irish Air Corps" with "auto-piloting, celestial navigation aids ... and many other 'new-fangled' ideas," its subsequent flight path to Europe passed directly over the Curragh, which wasn't rumour. Indeed, it was enough to drive any restive pilot crazy.

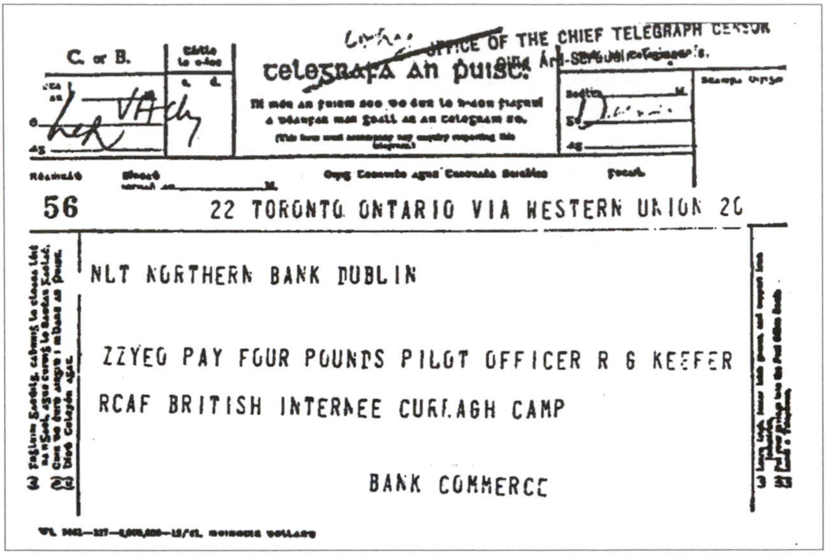

OFFICE OF THE CHIEF TELEGRAPH CENSOR

56 22 TORONTO ONTARIO VIA WESTERN UNION 20

NLT NORTHERN BANK DUBLIN

ZZYEO PAY FOUR POUNDS PILOT OFFICER R G KEEFER
RCAF BRITISH INTERNEE CURRAGH CAMP

BANK COMMERCE

Courtesy Irish Military Archives

receded to a large, flat expansion of twelve square miles, ideal for both horses (as the soil was rich in limestone) and military activities, which was why the British army had chosen the spot, before the British left for good in 1919. Now the challenge was to capture the belligerent internees escaping down the hill instead of the Fenians charging up it.

Although not today. Soon, the sun disappeared and the fog moved in, depriving him of much of the pleasure of his first trip to town. Errands aside, all that Keefer remembered of that trip was more sheep, bleating like lost ships in a fog bank. And no women.

A few weeks later Keefer met Susan Freeman, the daughter of a wealthy English shipping magnate who owned a large stud farm in eastern Kildare, some twenty miles from camp. He had little inkling that the first female to talk to him in several weeks would turn out to be neither Anglo nor Irish but Jewish. Or that she would occupy so much of his time in the months to follow. Indeed, he had thought little about women since the outbreak of war, so that when he started taking parole, his failure to meet any in the first month seemed only an extension of this lack of interest. There was a war on, after all, and his internment, together with all its inherent loneliness and

The two churches and the clock tower. Courtesy William Gibson

frustration, could only be justified in that context. So he chose other activities to begin with, like golf and betting on the horses at the nearby Curragh racetrack every Saturday. He was still concerned that if he met a woman he liked, it might lessen his desire to escape.

His girlfriend back in Montreal understood. She was an attractive nurse, the daughter of a Scottish businessman and a socialite who lived up on Mount Royal overlooking the French masses in Outremont and beyond, who, like him, had attended McGill. A daughter of the Protestant gentry. Although he hadn't seen her in over a year, he still expected to marry her, and probably would have before he left if he had thought they would have children right away, children who could preserve the family name should he and his brother die in service. But it never happened.

As for Susan, he met her at a foxhunt, the one activity that he had

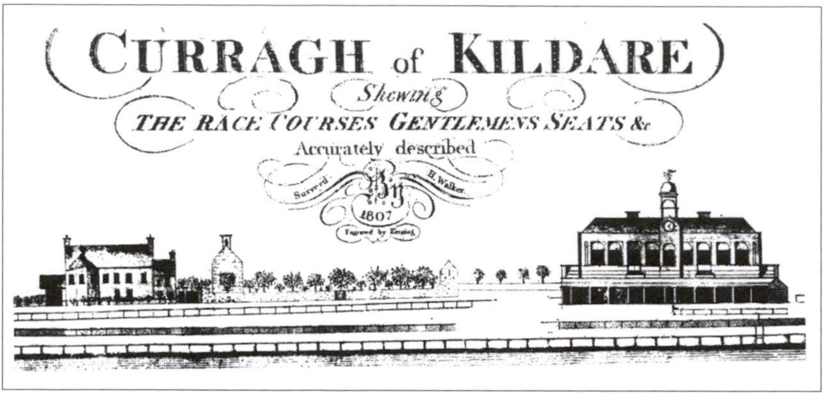

The race track. Courtesy Bill Gibson

initially sworn to avoid. It wasn't just any foxhunt though: it was the season's inauguration of the Kildare Hunt Club on the last Saturday of November 1941, held at Susan's parents' estate, Ardenaude, in Ballymore-Eustace. He didn't want to go at first, for he knew that attending a fox hunt would be seen by others as more evidence of his capitulation, but he allowed Jack to talk him into it.

"Christ, Bobby, do you want to go and meet some girls, or not? The Anglos are our only hope. First, no parole and now this. Celibacy's not good for a man of your age, you know that?"

"Well, I suppose there's no harm in meeting a few," he replied in a moment of weakness. "You're not going to write about me being on a horse, are you?"

"No. If and when the moment ever comes, I'll spare you that."

Because the hunt began early, they awoke at 05:30 hours so as to reach Ardenaude in time for breakfast. Their guide would be another English internee, Pilot Officer John Shaw, from Chelsea Park, in Essex, a fellow navigator who had negotiated Welply's landing that January. An affectionado of the hunt, and a good friend of the Freeman's, Shaw had helped them dress, laughing at the sight of the two Canadians in jodhpurs and spurs until they rebelled from embarrassment and insisted on wearing their tweeds. The three then hopped on their bicycles and headed east, for a two-hour ride through the midlands of Kildare.

It was his first sojurn beyond the plain, and with it came a whole new world for Keefer, one which he had hardly seen in rural England and was all the more astounded to find in Ireland – a world domi-

Pilot officer John Shaw

nated by the English or Anglo-Irish, the Planters, inheriters of thousands of acres of Protestant wealth on Ireland's best soil, the Pale.[16] It certainly wasn't what he or Jack expected to see – scenery which might have made a reporter wonder who held the power in Ireland, or why they were prisoners. Long ribbons of stone walls with borders of oak, copper and beach trees protecting large country estates built in the Georgian era for wealthy English aristocrats. Ring-necked pheasants, jackdaws, rooks and magpies, chirping away in

16 The original Planters, he told us, were English Protestants sent to Ireland during the reign of Henry VIII to assist in the colonization of the island, to "plant" the seeds of the Protestant faith. During the time of Elizabeth I, plantations were established in the literal sense, particularly in the area around Dublin, including Kildare, where the Horse Protestants settled. While those friendly to the internees may not have been Planters in the strict sense, they still owned, as late as World War II, most of the larger farms in the county.

late November as if they too would prefer to stay in Ireland, than brave the world outside. The sweet aroma of the countryside, before rounding a bend in the road and moving on to the next estate. In fact, Keefer passed so many that day that he lost count, estates with stables and carriage houses and servants and small concrete statutes and urns and ballistrades and fish ponds and towering willow trees and sprawling lawns and lavish country gardens. And none of them had addresses, either. They were too large to be described that way. They were known simply by title and county, yet they bridged counties as well, meandering from Kildare into Offaly, Meath, or Wicklow, and then back again.

Once more he was overwhelmed, not only by the level of wealth in a country that he had assumed was uniformly poor but by the surprises that kept popping up to greet him at a time when he was supposed to be bombing the hell out of Germany.

"Oh, yes. This one here ... That's Waddington-Brownstone's," Shaw explained as they pedalled up to greet the gatekeeper, "Midge's girlfriend lives here. Her father's 'a pound a year' man,[17] just like Freeman. Good friend of the Mitchells of Ballynure Eustache, joint master of the hunt and all. The Mitchells have a daughter too, Ann, beautiful girl. Welply and her didn't hit it off, though. There you go chaps, she's available. Wait a minute ... now that I think of it, I do believe that Major Mitchell was joint master of the Kildare Hunt Club last year, not this year. Forgive me. This year it's ... it's ... it's Major Claridge! Oh yes, this one here, a delightful little cottage," and so on, through Knockaulin, Kilcullen, Milemill, and Brannockstown as the three internees followed the River Liffey eastward. They even passed a dozen Germans along the way, cycling with typical Teutonic discipline in two rows of four each, strung out across the road just east of Kilcullen, apparently headed back to camp. The Irish had complained to the German embassy in Dublin about it, about how they always hogged the road, but the German internees continued to cycle that way, scaring everyone into the ditch, Irish and English

17 A pound a year man, he'd explain, was a well-heeled and influential businessman who offered goods or services to the British government for the token sum of one pound sterling per annum. Several in Kildare at that time fell into the category, although few would have announced it publicly for fear of offending Dublin.

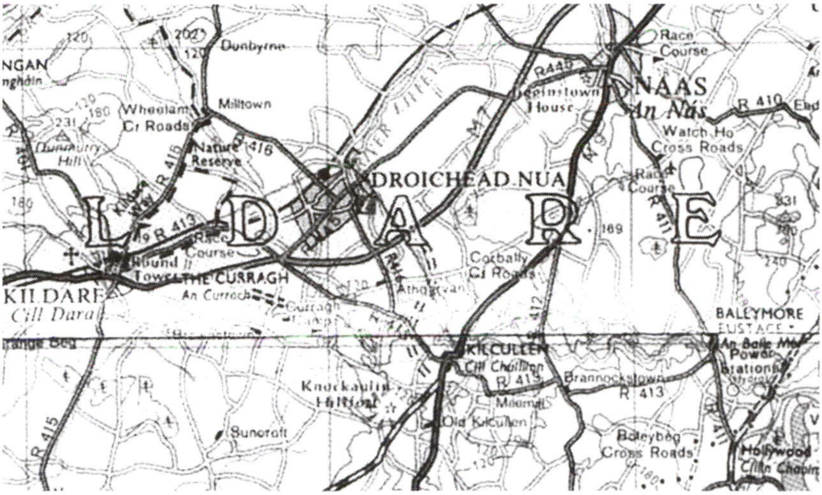

Roughly the parole boundary, including the Curragh, Droichead Nua (also known as Newbridge) and Naas.

alike, making their presence known. It had led to a brawl between the RAF and German internees one afternoon, just outside camp on the road to Newbridge (Droichead Nua), when both sides refused to move, claiming that they were each on the proper side of the road which they would have been, had they each been home. This time, however, the three RAF officers smiled nervously and ignored the insults. No one was going to ruin this day.

At 08:00 hours sharp they arrived at Ardenaude. There was a traffic jam of horse-drawn carriages, Addlers, Model T's and A's, and weekend horse trailers from Dublin, left hurriedly beside the gatekeeper's house by the hunt's hungry participants who had set off up the hill for breakfast. The two Canadians pushed their bicycles through the main gate and headed up the hill.

T. Spencer Freeman, owner of Ardenaude, had founded the Irish Sweepstakes in the late 1930s. He had purchased Ardenaude as an investment, and with it a stud farm of 650 acres that was maintained with a staff of thirty Irish labourers and six servants so that his young wife, Toggs, didn't have to contend with the rigours of country living.

"The old girl keeps a pretty busy social schedule," Shaw explained.

"Why 'Toggs'?"

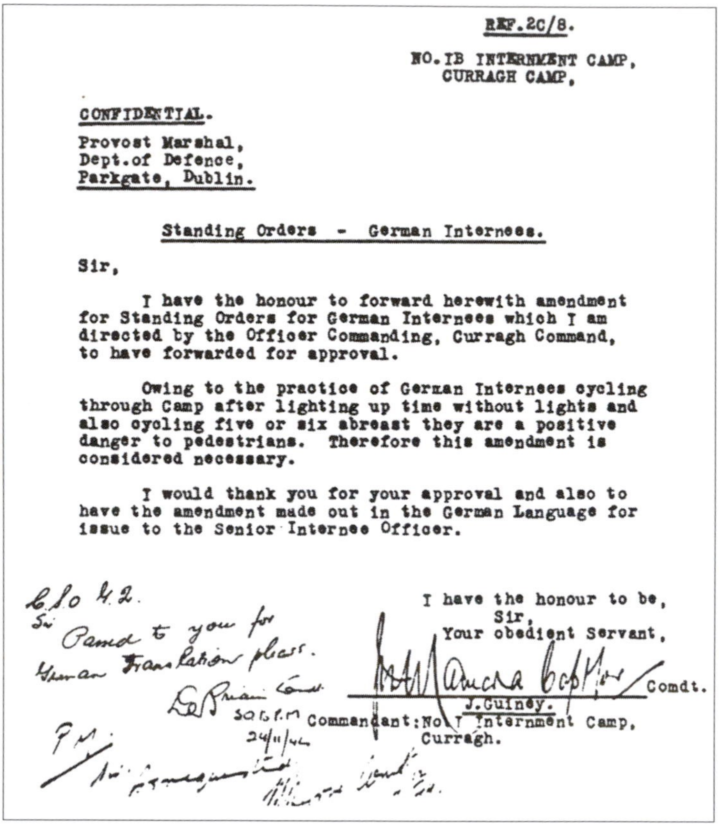

REF.2C/8.

NO.IB INTERNMENT CAMP,
CURRAGH CAMP,

CONFIDENTIAL.

Provost Marshal,
Dept.of Defence,
Parkgate, Dublin.

Standing Orders - German Internees.

Sir,

I have the honour to forward herewith amendment for Standing Orders for German Internees which I am directed by the Officer Commanding, Curragh Command, to have forwarded for approval.

Owing to the practice of German Internees cycling through Camp after lighting up time without lights and also cycling five or six abreast they are a positive danger to pedestrians. Therefore this amendment is considered necessary.

I would thank you for your approval and also to have the amendment made out in the German Language for issue to the Senior Internee Officer.

I have the honour to be,
Sir,
Your obedient Servant,

Comdt.
J.Guiney.
Commandant:No.I Internment Camp,
Curragh.

Passed to you for German Translation please.

24/11/44

Complaints about Germans cycling, four, five, or even six abreast, like other aspects of his story, were substantiated following the release in the early 1980s of official reports like this one. Not that we didn't believe this particular claim, only that it seemed too predictable to be true. Courtesy Irish Military Archives.

"You'll know when you see her."

A moment later a woman appeared on horseback. She wore a scarlet waist coat, a black calf-length skirt, white gloves, mahogany-tipped riding boots with the family seal, and a feathered hat. "Hello, John, smashing to see you," she said, dismounting and crossing over to where Shaw stood, offering her cheek. She appeared to be in her early forties, and was quite attractive.

"Toggs, I'd like you to meet Bobby Keefer, the Canadian pilot I've been telling you about." The woman smiled at Keefer, taking him by the hand.

Ardenaude, county Kildare.

"And this is his navigator, Jack Calder, the reporter."

"My goodness, Susan and Bridget will be delighted to meet you two," she said approvingly. "I had no idea that your two friends were so good looking, John. You should have brought them by much sooner. Which one is the one who refused parole ... Bobby, is it? Ah, the girls will be so impressed. And Jack, the famous Canadian reporter. Yes, we've heard about you. When will we get to see our pictures in the press? Wouldn't that be fun! Come, come, now, breakfast is waiting."

As they walked up the hill, they passed a tennis court with its rear net draped in ivy, a badminton court, and an overgrown croquet pitch. Behind the croquet pitch was another garden, also overgrown, with purple heather extending out over the lawn.

"Now that the girls have grown up, I'm afraid they're no longer interested in that sort of thing," Toggs said apologetically, gesturing towards the garden and stopping to snap an encroaching bough from a laurel bush. "And the staff haven't been the same since Spencer left. Turnover's dreadful. Hard to find loyal Irish servants, I'm afraid, not like in the old days. Not as motivated, I fear, and half of them spy on us, work for the bloody government.[18] And the dear girls, well, all they ever do is talk about dancing, parties, and men. I do hope they won't bore you."

Ardenaude, county Kildare.

When they had reached the top of the hill, she directed them to a building off to the side.

"John, you and your friends go and wash up before breakfast … no, no, not that door, the servants' door, over there … Patrick!" A man in black had appeared at the carriage house door. "Give these boys a scrub before breakfast, please. I don't want the girls seeing them like that." She turned and headed off towards Ardenaude's front entrance.

In the servants' quarters they were provided with scented soap and hand linen and then directed through the garden, along a path of moss and bracken beneath the Corinthian porch to the front garden. From there they were shown into the house itself, where they turned down this hall and that until they were in the main foyer, where they met Toggs.

"Nothing like a man who smells nice in the morning," she said approvingly, once more grabbing Keefer by the hand. She appeared confident, warm, and friendly, whereas he felt uncomfortable wear-

18 While no documents at the archives record specific examples of such spying, a number refer to information regarding members of the gentry in circumstances suggesting that the source must have been a "domestic" spy. This is also consistent with the suspicions, many years later, of the gentry members themselves. As will become apparent, G2, the intelligence arm of the Irish Army, was very adept at tracking the movements of the internees while on parole.

ing the cologne Patrick had splashed on him moments earlier. "Now, come over here, Bobby. I would like you two to meet my daughters. Susan this is Bobby, or should I say Pilot Officer Keefer of the RAF, the Canadian pilot at the Curragh that I was telling you about my dear, you know, the one who refused parole. And Bridget, this is Jack, the reporter." Keefer had stopped listening, cringing at the continuing reference to his discontinued protest.

"I know Ma, I know. Don't embarrass him. He's obviously a modest man and not accustomed to such fame." Susan, who seemed more subdued than her mother, grabbed Keefer's hand. Taken aback, he mumbled a quick greeting that no one could possibly have heard, or understood. "Relax, Pilot Officer Keefer, I won't bite you," she replied. "If you're not afraid of the Germans, then you shouldn't be afraid of me!"

She had spoken quietly, holding onto his hand longer than he would have expected. To make matters worse, she stood so close to him that a strand of her hair was touching his cheek, carrying the scent of the perfume she wore on the collar of her white-laced blouse. Her eyes studied his mouth, before returning to his face.

"You have a lovely home, here, Miss Freeman," he said awkwardly, gazing about the foyer and its adjoining parlor where most of the guests, in red coats, black coats, swallowtails, and bowlers,

Ardenaude's humble servants' quarters.

had congregated.

"Thank you, Pilot Officer Keefer, at least I could hear that. Now come and get some breakfast. Have you been aboard a horse before?"

"Not really, but I'll try," he said, lying. He had ridden horses before, he just wasn't very good at it.

"That's better," she replied. At least the mumbling had stopped.

Susan's long hair was tied at the top of her head, with strands falling down the side of her face. It was dark brown with a hint of auburn, the same colour as the rosewood sideboard on which he had leaned, nervously, when she had taken him by the hand so unashamedly, so boldly. It startled him. He could have gone through a dozen dates with most girls without being grabbed like that. As they stood by the table waiting to sit down, he tried to keep his eyes off the rest of her, or at least the upper half, shaped to prominence by the half-opened coat and gold-brocaded belt tied snugly about her waist. She was beautiful, more beautiful than her mother or her sister. From time to time he had the courage to look at her directly: her forehead was flat and smooth, her nose delicate, her eyebrows thick, and her skin dark rather than pale, yet it was her eyes that finally held him, finally calmed him down long enough that he could at least look at her.

The hunt commenced immediately after breakfast at which point he was faced with a whole new set of obstacles, for riding a horse did not come as naturally as flying a plane.

"Just climb aboard the Byerly Turk here," said Susan, walking beside him as the hunt's participants gathered in the front garden. "Aragon, his name is. Let me help you. Put your hand on the saddle and hoist ... ah, let's try again, shall we," she said, as he struggled to perch his large frame on top of the small English saddle.

"That's better. Now tuck in your trousers, keep your knees pressed to the horse, and sit upright ... there you go. Stay as close to me as you can, Pilot Officer Keefer. We'll be heading out to the covert where the foxes are, beyond that hill."

Susan was now sitting bolt upright in the saddle, cantering directly ahead of him, looking out towards the sun and the distant hills. Her hair, no longer restrained by her barrett, fell on her shoulders; her square back smooth against the waist-coat down to her hips; and her pale-grey skirt, tight-fitting, with the shape of her thighs and hips plainly visible. She looked both feminine and strong, a combination he found both attractive and intimidating, and not at all like his Montreal girlfriend.

"The hounds'll be waiting until the first covert's drawn and then the master will yell 'tally-ho' and all hell will break loose. So stay close and don't get up with the hounds or they'll call you a thruster and we don't want that, now do we."

"No, Miss Freeman."

"There will be plenty of hounds, maybe twenty couple or so, and some of them riot, so you'll have to stay clear of that man over there." Susan had moved over to him so that the two horses were side by side, and touched him on the arm, directing his attention to a man on horseback waiting by the triple carriage-doors. "That's the whipper-in."

"The whipper-in."

"The whipper-in, and his job is to bring the hounds back to the pack, and he'll be mighty upset if you get in his way. Now we might have to do some jumping of hedges and walls and that sort of thing, but don't worry, you just watch me. If there's any problem, I'll come down and dig you out of the primrose. Come along Pilot Officer Keefer, the foxes are waiting."

As they set off after the others, she was careful to remain beside him while they trotted past the carriage-house, beyond a small stone quarry, and then along a trail into the forest.

"What's a thruster?" he asked, as she was admiring his handsome, worried face.

"Someone who keeps too close to the hounds," she replied. "So stay close to me."

"Well, shucks, ma'am, if you're not up there eatin' their dust, how can you expect to rope 'em." He startled her with his finest Gene Autry Western accent, of which he was particularly proud.

"I thought you were Canadian." She smiled. "So you do have a sense of humour." The two remained together, throughout the hunt and at the dinner that followed.

In fact she quite liked what she saw, the polite shyness without the affected mannerisms of the English officers she so despised. She had told her mother that and her mother had promised to find her someone different, an Australian, or New Zealander, or an American maybe. And now she had found the next best thing.

So they had meandered together in and out of the oak and beech trees, well back of the others as Keefer struggled to keep up, steeling himself for the challenges ahead. Dancing after dinner in Ardenaude's expansive ballroom to Jimmy Dunny and his ensemble, a jazz band of sorts, with a brass section from England and a string section from Claire, playing an oddball collection of American classics of the day like "Six Flats Unfinished," "The Jersey Bounce," and "Don't Be That Way." And as they danced, they would talk, of theatre shows, sports, and friends back home, of music and fashion. But never of war.

5

CELEBRATING PEARL HARBOR

Arriving back at camp that night it must have occurred to Jack Calder that the story he had just finished about his and Keefer's misadventures in county Clare – a jaunty 3,000 word account entitled "I Flew into Trouble" (picking up on the "I Bombed the *Gneisenau*" theme) – might have to be revised. Not that he could do that, since he had sent it off with his brother the day before – one of the many odd features of their internment being that they were regularly visited by friends, relatives, and fellow squadron members who would come down from the North, wondering what all the fuss was about. The problem was that the more time he spent on parole, the less he realized he knew about the place. Next time, he decided, he'd do a little more research before shooting his mouth off.

Not that Jack shouldn't be forgiven for picking up on the stereotypes and ignoring, for the time being, the complexities. Or for the fact that life in the Pale wasn't exactly what he expected, since he hadn't exactly expected to be there in the first place. No, the truth was that Ireland in the 1940s was every bit as complex and misunderstood as it had been before the war (and would be after), especially to one trained to view the world and the Commonwealth through their eyes, through English eyes. Through the eyes of Disraeli, say, the former British prime minister who, in the debate over Home Rule fifty years earlier, had described the ungrateful scoundrels as "scum condensed of Irish bog, as a reckless indolent, superstitious race daring to run roughshod over England, haughty, yet still Imperial England," a

Jack with his typewriter. Although Girdlestone always claimed that Jack got the typewriter from Colonel McNally on account of what was to follow, the more likely source was a well-known merchant in Newbridge (Droichead Nua). Courtesy Bruce Girdlestone

view not dissimilar from what was being espoused in gentler language by the editorial boards of the Manchester *Guardian* and the *Times* in Jack's time: the breakaway ruffians of the South who had abandoned the Crown in its hour of need.

He certainly hadn't observed that about the Irish so far, and not just because he considered himself half-Irish. As a reporter he was inclined to investigate first and then draw his own conclusions.

On the other hand, he was aware, based on the observations of the great H.J. Morton, the well-known English travel writer of the 1930s whose books were popular in Canada and around the Commonwealth, that the place was a little odd, "strange" as Keefer put it or, as Morton himself said, "that there was something drawn out of the earth, out of the water, in the streams and grass and flowers ... of the sky; that if a man could ever hear it, he would know all there was to know about Ireland." Romantic Ireland, Jack thought, though he preferred to describe the place as "lulling," especially in contrast to what he and his friend had just been through on the far side of the Irish Sea.

And he did.

But to understand the rest – that took some effort – the legends, the myths, the reasons for the superstition and indolence; the role of religion, of course, and Irish neutrality in its proper context. For America was neutral too, and the English weren't jumping all over them, at least not to the extent they were the Irish. And then there was the historical thing, beginning with the arrival in the first millennium of Mesolithic hunters in skin coracles (not unlike the exterior fabric of a Wellington), the passage tombs of county Meath, and beyond. Why it was so important to keep reminding everyone about *their* kings, listening, with fascination, one afternoon in the cold rain as Colonel McNally recounted the legend of the Curragh, of St Brigid and her ears – how in 480 AD the patron saint of Kildare had created the plain by ridding the king of Leinster, one of four ancient kingdoms, of a deformed ear in return for all the land her cloak could cover; of how the plain had been awarded to the Fitzgeralds of Pemgroke as recompense for service to the Crown, when one Irish king, Ternan O'Rourke, abducted the wife, Dervogilla, of another Irish king, Dermot MacMurrogh, as a favour to the English, leading to a long history of domination in the Pale. Rather than the English kings which Jack knew about, like Henry II, or James I, or King William. Or Cromwell, who wasn't really a king, but sure got a lot of coverage over there.

He had had to explain the typewriter somehow, so he told the colonel that he was studying Irish history and planned to write a book about it one day. The colonel seemed pleased. Of course, he felt obliged to include a passing reference to that in his article – Irish Kings, that is – but like most careful writers Jack didn't want to say any more until he had figured it out. And that would take some time.

So, in sum, Jack was a bit worried about what he had just written and sent off for distribution on the wire. Not so much because of the possible reactions of his colleagues – for the public had a right to know, after all – or even that his subject matter was less important than, say, taking out a German pocket battleship. It was more a matter of failing to meet his own high standards. He'd have to correct that next time.

When Pilot Officer Roland (Bud) Wolfe, the Curragh's first American internee, arrived in camp in the first week of December 1941, it would have been newsworthy in any setting. The sudden, unexpected appearance of an American could do that – particularly in neutral Ireland, in the week prior to Pearl Harbor, when indications were already strong that Roosevelt would announce America's involvement in the war. Everyone knew that de Valera, born in Manhattan to a mother who had emigrated there and saved from execution in an English jail by the Americans, would follow suit. He'd have no choice. His benefactors would see to it.

"Yes lads, I agree," declared the bar officer from his usual post behind the bar, barely containing his excitement. "It won't be long now."

They had just analyzed the situation carefully, dissecting the various editorials on the topic, not just in the English and Canadian papers but in the *New York Times* and *Washington Post* as well; considered, with Jack's help, the views of interested reporters, leaders, and diplomats, before finally determining that this was it, that the camp would be closed and they'd all be released.

"Not long at all, lads, not long at all," agreed Midgely.

"It has been pleasant in a way, though hasn't it?" added Shaw.

"Yes, I suppose," reflected Welply, seated on the stool to his right.

"Still, it's time we got back," said the bar officer, the most impatient among them.

"Indeed, it is," they agreed, "indeed it is."

"A toast then. To Belfast!"

"To Belfast," they all replied, not just the English officers but all of them.

For a full contingent of officers had come out to greet the American that night, ten in total, including, apart from the English officers, the two Poles, PO Karniewski and PO Baranowksi, who had arrived after separate crashes in March and April of that year – in the former's case following a ditch in Dublin Bay where he was rescued and then feted by the predominately Protestant members of the Dublin Yacht Club before being carted off to the Curragh, and in the latter's, under suspicious circumstances, to say the least, in good weather with sixty gallons of gas in his tank. Another new arrival that week, Sub-Lieutenant Bruce Girdlesone, from New Zealand, a volunteer pilot with the Royal Navy Voluntary Reserve was there,

Sub-lieutenant Bruce M. Girdlestone.

Flight Lieutenant J.L. Ward.

PO Stanley Karniewski. No photograph or sketch of Baranowski, whom they called the Baron, survived the war, adding to his shadowy reputation.

as was Maurice Remy, a fighter pilot from the Free French forces who had landed in June and was lionized by his colleagues for having downed a Heinkell III near Kimachthomas in county Waterford; and, finally, Flight Lieutenant J.L. Ward, an Englishman by birth who had moved to Vancouver before the war, who had arrived that January after his Whitley bomber had crashed at Cockaveney, County Donegal. Technically their commanding officer by seniority (and more English than the English), Ward had wrangled a "special parole" that permitted him to live off the base on a Protestant stud farm with his wife, who had flown over from London to join him. His only obligation was to check in every few days with Captain Fitzpatrick at the police hut – not unlike the arrangements in neutral Portugal or Spain where the internees were obliged to check in with the local police chief every Friday afternoon, assuming they could find him.

As for Wolfe – who had crashed the day before when his Spitfire had seized up over the Lough Foyle – he didn't really know what to think. No, that's not true. His reaction to all of this was similar to Keefer and Calder's, only more so.

"You mean to tell me that those tankers were bound for Ireland all along." At the time of the crash, Wolfe had been flying escort for a convoy which he had assumed was headed to Glasgow. "What in hell do these assholes need oil for? They're not the ones fighting."

The English government, in an attempt to sway public opinion in

Pilot Officer
Roland L (Bud) Wolfe.

Ireland, had opted for a partial easing of the embargo that fall, following the failure of renewed negotiations over control of the ports.

"It's a bloody crime, isn't it Wolfe," replied the bar officer, popping another bottle. "Now you do understand this business about the oath, don't you? It's very important. You see, you sign a parole slip like this one here." Covington reached behind him for a cancelled parole slip that Colonel McNally had provided, worried as everyone was with the repercussions of an American in camp.

"You sign right here, see Wolfe, on the dotted line."

"Yeah, yeah, yeah. Listen Covington. Don't expect me to hang around this dump any longer than I have to."

The American looked up at the others, who were viewing him with suspicion.

"What are you jerks looking at?"

Wolfe, who had volunteered for the RAF for slightly different reasons than the two Canadians – more as a vocation, really, than a crusade – was clearly unhappy with the recent turn of events. Keefer had met a lot of men like him in Saskatchewan, since Americans preferred training with the RCAF as they were not required to swear allegiance to the Crown. "Just gimme the stick, baby, and lemme fly."

In fact, unknown to any of them, Wolfe had another form of parole in mind – in the United States – which he had just presented to the government via the American minister to Eire that morning. Not that any of them would have been surprised if Wolfe had bolted before de Valera had had a chance to read it.

The euphoria continued well into the night, well past last call, with the inner cabinet and the supply of Guinness once more called into account. And by night's end each was convinced that if he wasn't back at his base by the end of the week, then at least he'd be back for Christmas, or certainly before the New Year.

If the Yanks had anything to say about it.

The following Saturday the Japanese bombed Pearl Harbor, and by then they had all started packing. When the news was announced on the wireless (in Ireland on the next day, 8 December 1941)

Prime Minister deValera
December 2nd, 1941
[Copy-typed]

Sir,

I have the honour to request that I be sent home on parole as I am a citizen of the United States of America.

I would give parole not to take up arms against any of the axis powers.

I was a member of the Eagle Squadron whose members do not swear allegiance to the King or his Country. They are, more or less, just working for hire.

If I can do this I shall be able to continue with my civilian flying and have a profession after the war. If I remain here there will be nothing I can make a living at when this war is over.

Sir, your obedient servant,

R. L. Wolfe

Courtesy Irish Military Archives

Keefer was in the officers' bar, talking over old times with Jack, girls, (most notably their recent letters from Dode and Blink, the redhead and brunette they'd charmed at Uncle Wilks' estate in Surrey) and the fact that he now knew what the others didn't – that Jack had returned briefly to his former occupation and that sooner or later they'd all find out by seeing their names in print, unless they were released by then, which in Keefer's view would be best for all concerned. Especially Jack. The radio was tuned to the BBC, as it usually was, with Remy there beside it. Newspapers and magazines were strewn about. They had just returned from evening mass at the Protestant chapel where Reverend Flint, a retired Anglican Army chaplain, had delivered a cautious sermon predicting the end of the war, as he did most Sunday nights. The service had been attended mostly by RAF internees, including the NCOs, but also by a handful of German internees. And then suddenly the announcement came – the Japs had attacked Pearl Harbor, a place that no one except Jack had ever heard of. Not to worry. If blood was spilt on American soil, their time was up. The celebration began.

"Well that's it, lads. Looks like the end," announced the bar officer, reaching once more towards the inner cabinet, barely able to contain himself. He had spent a full year there wasting his life. Now

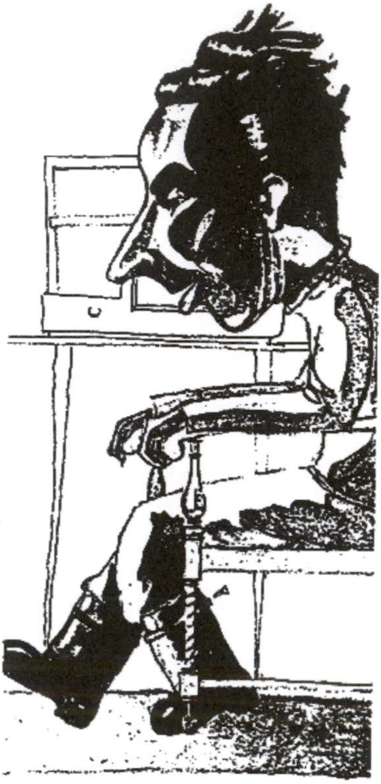

Pilot Officer Maurice Remy,
listening to the wireless.
Courtesy Bruce Girdlestone

he could get on with the rest of it. Fortunately the diplomats had been by the previous day, arriving in a large black Addler, two stiff bodyguards in long greatcoats carrying cardboard boxes through the main gate – Sir John Maffey, John Kearney (the Canadian high commissioner), and Wing Commander Begg, the RAF's attache in the lead. David Gray, the American minister, had been called to Dublin on urgent business, a promising sign.

"Bang on, Bud, old boy," chimed Pilot Officer Welply, mug in hand, slapping the dazed American on the back. "Bang on, old boy!"

"Damn right," replied the excited American, experiencing a sudden change of heart. "MacCarthur'll blow those assholes from Berlin to Budapest."

"The Americans will come and free us," added Stanley Karniewski, the younger of the two Poles, who had lost his homeland to the Germans three years earlier. He was in tears.

From left, Covington in his usual post, Girdlestone, Karniewski, Ward (given the occasion) and Bud Wolfe, closest to the camera. Courtesy Bruce Girdlestone

"Ah mon dieu, les Americans s'arrivent. Les Americans s'arrivent," declared their French colleague, Maurice Remy. His confidence in the Brits had lagged, ever since Dunkirk.

By the time the bar finally closed that night and the singing had stopped, it had been a celebration to rival the one that spring over the sinking of the *Bismarck*, fueled by the nightly choruses of "Deutschland Uber Alles" that wafted over the camp walls. Indeed, as Keefer recalled years later, he and his fellow officers must have been the only people in the allied world who actually celebrated that night.

The following day, as spirits remained high, Keefer, Calder, Covington, Welply, and Midgely took the American out on parole. Their destination was the camp links, followed by a few pints and dinner at Osberstown House, a popular hotel and meeting place along the river Liffey, north of Naas. Osberstown House was owned by the Lawlor's, a well-known Catholic family who catered for many of the hotels in Dublin. It was famous for its dances, particularly in the hunting season (which in Kildare stretched over the autumn and winter months) and for the fact that Germans often attended, though rarely on the same night. Keefer figured he was bound to run into Susan there, something he had been avoiding, given his imminent departure.

"So have you seen one of these lads?" Captain Fitzpatrick asked at the police hut as he held a parole slip in front of their newest internee, Pilot Officer Wolfe.

Wolfe looked disgusted.

"So you're the American pilot, are you?"

In fact the American pilot was more afraid than disgusted. What if the guard said something to the others? It would be one thing if the Irish found out about his letter to de Valera; it would be quite another if these guys did. He didn't want them to think him a shirker, especially now.

"This is known as a parole slip, Wolfe. You're to sign on the dotted line. The rule is that you're allowed out until 02:00 hours. If you come back before that, you've got to get your slip stamped." He paused to show them the stamp for, in Covington's case, the umpteenth time. "If you go out after that, you've got to get a new one. Signed. Got it?"

"You'll get it soon enough, you Irish bastard."

"Acch, Midgely, such language. I haven't heard that language since the Troubles."

"Come a little closer and I'll show you the Troubles."

The internees laughed, now accustomed to the banter. And anyway, they were going to be released within a matter of days.

"You know how Colonel McNally feels about your language," said Captain Fitzpatrick, suddenly serious. "He might cancel your parole."

"Let him try, you Italian scum." It was Welply's turn this time, Ireland, like Italy, being seen by most of them as little more than a pawn of the Third Reich. After all, there was nothing the Irish could do to them now.

The guard sneered. "Call me Irish if you might but not Italian." He noted the exuberant mood. "I bet you lads are on about Pearl Harbor. Is that it?"

Wolfe flinched. The guard looked directly at him and then away.

On the links they divided into threesomes at the first hole. Wolfe and the two Canadians were in the first group, with the English in the second, a mini Ryder Cup. Their excitement and anticipation carried them through the first few holes. By the third, Wolfe had decided to trust his new friends with his secret. As fellow volunteers, as fellow outsiders, they shared a bond. He told them about the letter.

"I think that's beyond the ten mile limit," replied Keefer, aghast.

"Maybe, Bobby, but I didn't come 7,000 miles to play politics."

"What else did you say?" asked Jack.

"I told him I was working for hire and that if de Valera let me take parole back home, I wouldn't come back."

"Come back?"

There was silence as Wolfe bent down to tee his ball up on a tuft of grass. He looked at the two Canadians and shrugged. "That was before."

"Still, Bud, do you know what they'd do if they found out? Your career would be ruined."

"Maybe, Jack. But why fight if they treat us like this?"

"Who? The Irish or the RAF?"

"Shut up, Jack. They could always lock us up," added Keefer.

"Or they could just release us, Bobby," countered the American. "It's bullshit, and I can't believe you guys are falling for it."

Wolfe then slapped a screaming slice into the gorse, watching as the sheep scrambled for cover. That made him feel better.

At Osberstown House after the match, the mood was still upbeat, with the various participants settling into their pints, tallying their cards, arguing over who owed what, and planning their first few weeks of freedom. After the North Americans had paid up, the two Canadians announced that they would go immediately to London, to the Regent, to blow off a little steam and pick up where they left off, back at the 103 if their squadron would have them. Though Jack's brother had mentioned that he had heard that former POWs could pretty much write their own tickets, so maybe they'd have a choice – though no one was sure if the policy would apply to them. They weren't really prisoners of war in the traditional sense. Covington said that he had received a letter from a former internee who had escaped in June that year, Paul Mayhew, who had told him that the RAF had let him write his own ticket, but then he was well connected. Mayhew, who had arrived at the Curragh with Covington a year earlier, had escaped over the wall with two others that July during the Irish derby. His father, Sir Basil, a high-ranking

English diplomat, had arranged the escape, enlisting the assistance of MI9, the British intelligence agency, who had sent in a handful of agents to cut the wire and help the escapees to the North.[19] The Irish kicked up a fuss, however, and the RAF had forbidden any similar attempts in the future. Still, just talking about escaping made their release seem all the more certain.

A short while later Bridgit Lawlor entered the bar. Mrs Lawlor, in her mid forties, had two sons, Jim and Tom, who ran Osberstown's. They were present and introductions followed. As Keefer would recall years later, the two brothers, in their early twenties, became good friends of his and Jack's, even going so far as to resurrect their nicknames – Jackie FitzCalder and Babby O'Keefer – the following summer on the golf course, appreciating the fact that, unlike the English internees, the two Canadians had nothing against them because of their religion. They also met Josephine, a tall, attractive woman, who kept the bar. She was fluent in three languages, English, Irish and German, the perfect barmaid for neutral Eire in the 1940s. While Keefer clearly remembered her, his eyes were nevertheless trained on the door waiting for Susan to show up. While he didn't expect her to appear, nor really want her to, given the fact that he'd be leaving in just a matter of days, he wouldn't have minded if she had.

By the following week, the mood in the camp had changed. The internees on the RAF side were getting a little worried. Beyond their own situation, they were obviously delighted that the Yanks would now be coming to Europe, knowing how the English needed their help. But they also knew that, as of the Fall of 1941, while Germany might not have been winning the war, neither were they. So whatever relief was felt from Pearl Harbor was tempered by a growing realization that their services were desperately needed.

And nowhere was that realization more apparent than with the camp's newest arrival, Pilot Officer Roland Wolfe.

On Friday morning that week – December 13th as luck would have it – Wolfe returned early from parole. He had been expecting a visitor from the North, a fellow member of the No.3 Eagle

19 Paul Mayhew's brother, Christopher, served as England's secretary of defence in the 1960s.

Jim Lawlor, Keefer, Tom Lawlor, Calder, June 1942. Courtesy Bruce Girdlestone

Oberstom House Hotel. Courtesy Vi Lawlor

Osberstown House Hotel, Naas, Co. Kildare

Squadron, Flying Officer Johnny Jackson. When Jackson arrived shortly before 16:00 hours, Wolfe greeted him at the main gate. "Cheeri-bloody-o! Johnny, great to see ya," he said. The two friends hugged, slapping each other on the back.

Passing through the police hut, Wolfe then signed a new parole form. The plan was to head off to Naas with his friend for dinner and a few pints; he filled in 02:00 just in case. After dinner the two returned to camp early, at about 22:30. When they came to the police hut afer passing through the main gate, they discovered that Captain Fitzpatrick was nowhere to be seen. Maybe this Friday the 13th would turn out to be Wolfe's lucky day.

Thinking it through, the American decided that he had been presented with a glorious opportunity. The reasoning was convoluted: if Fitzpatrick had been present, he would have cancelled Wolfe's form, which would have restricted the American to escaping from inside the camp. Escaping from the outside was forbidden, obviously, which was the whole point of the system. With Fitzpatrick absent, however, Wolfe could cancel the old form and leave it on the desk. He could then fill out a new form, cancel it as well, and leave it on the desk too. Then if the sentries failed to ask him for the new form on the way out (if they did, he'd claim he'd forgotten, before dashing back to the hut to get still another), he'd find himself outside the camp, without technically having given his oath or promise to return – that is, as if he was still inside. He would then be free to escape.

It was an interpretation of the oath based more on the letter of the law than its spirit. It was also one they had discussed on the golf course the previous week, after someone had noted Fitzpatrick's occasional absence from his post.

Wolfe asked his friend Jackson what he thought. Jackson had no idea what to think.

Wolfe returned to the empty hut, carefully shielding himself from the spotlight. He grabbed a form from Fitzpatrick's desk and signed it. He then headed out with Jackson to the main gate.

"Your form, please, Wolfe," asked Corporal Reilly, the younger of the two turnkeys usually stationed there. The younger officers tended to be more polite, though quicker to temper.

"Yes sir, Corporal," said Wolfe, nervously presenting the form, duly signed. The guard looked suspiciously at the form, then at the two flyers. Wolfe cursed under his breath – he should have been ruder! It was properly signed, however. So now all that Wolfe needed to do was to get back inside.

"A little late to be headin' out," noted Corporal Smeaton suspiciously. Smeaton was the older of the two turnkeys at the main gate, seated to Reilly's left. He was leaning forward, studying the form.

"Just a nightcap Corporal," replied Wolfe, "just a nightcap."

Yeah, thought Jackson, in Belfast.

"Fine. Go ahead lads," said Smeaton. So far so good. Wolfe then began rubbing his hands together, raising them to his mouth and blowing, a moment of genius, really. "Bloody cold out, isn't it Corporal," he said, shivering. "Mind if I go back and get some gloves?" As the son of a Nebraska farmer, maybe Wolfe did find the damp Irish nights uncomfortable. Regardless, it worked. The two nodded again in agreement. Leaving Jackson at the main gate, Wolfe quickly returned to his hut and grabbed his gloves. He then made his way back to the parole hut. His heart was pounding. What if Fitzpatrick had returned?

Bud Wolfe. Courtesy Bruce Girdlestone

No Fitzpatrick. Wolfe quickly reached over the counter, grabbed the stamp on Fitzpatrick's desk, and stamped his form "Cancelled." He then left it in the tray for all to see.

He ran back out to the main gate. Much to his delight, Corporals Reilly and Smeaton merely turned away, allowing the American to proceed. Finally, after two weeks of this nonsense, he was free!

Wolfe and Jackson walked two miles to the nearest bus, which took them to Dublin. There they stayed at a hotel and caught the 10:00 train to Belfast the following morning, reporting that afternoon to their unit in Eglinton, county Londonderry. Having left his last slip on the desk, cancelled, and having then passed through the main gate that day without signing a third, Wolfe had escaped. Legitimately. And there wasn't a thing the Irish could do about it.

"Avez-vous des moules?" asked Maurice Remy, the Curragh's only French internee, of Albert, the head waiter at Jammet's, Dublin's only French restaurant.

"Des moules, Monsieur, en Irlande?" Albert replied in the snotty, contemptuous tone used exclusively by head waiters, especially French ones. "Impossible." Little did the little man know. A shipment of Atlantic lobsters to the Curragh had just been arranged for the following week, courtesy of the Canadian High Commissioner Mr Kearney and RAF Ferry Command.

"Des homards? Des langoustines?" Remy, who missed his homeland, craved fresh seafood.

"Non, je regrette, mais il n'y en a pas."

"C'est étrange, Albert – nous avons des moules et des homards chez nous."

The headwaiter at Jammet's smirked. He was getting impatient, but didn't dare show his impatience in the presence of Maurice Remy, the darling and adopted son of the restaurant's owner, Louis Jammet, who had given Remy a standing invitation to dine at Dublin's most famous French restaurant. With or without his colleagues, though his colleagues were always welcome, colleagues whom, like Remy, Jammet had read about in the local press, either their landings or their daring escapes. This was Jammet's way of showing his gratitude, and of making it clear that, contrary to what many of them believed, not everyone in Dublin in 1941 was prepared to concede his homeland to the Third Reich.

Remy, whose father had died in World War I, and who had been abandoned by his mother on the streets of Paris, was more than happy to be adopted by such a distinguished Dublin restaurateur. For one thing, it gave him a leg up on the English internees. If and when they visited the Irish capital, they were seldom invited anywhere.

"Avez-vous choisi, M. Remy?" Albert had taken a new tack now, chin toward the table and eyes on the chandelier, lest he pay Remy the compliment of looking at him directly.

"Non, Albert, demandez mes compagnons, si'l vous plâit," Remy replied just as dismissively, with a flick of his head in the direction of his two Canadian colleagues who had accompanied the French internee on their first Dublin parole. After a month of golfing and frolicking with the daughters of the Protestant gentry, even Keefer felt it was time for a change. (Had they known what was going on at the Curragh, on the other hand, they would probably have remained behind.)

"Messieurs?"

Keefer, proud of his ability to understand at least some of

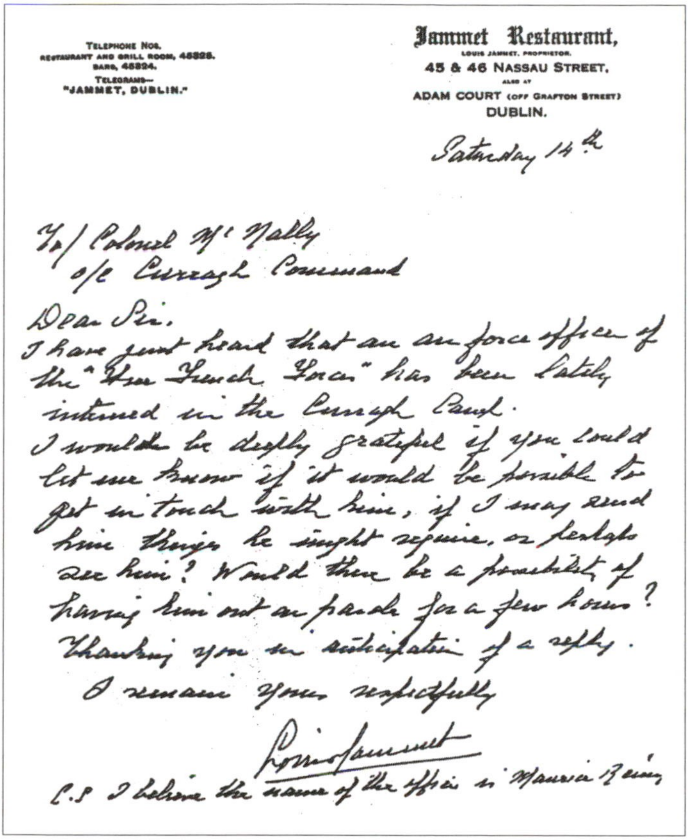

Telephone Nos.
Restaurant and Grill Room, 45828.
Bars, 45824.
Telegrams—
"Jammet, Dublin."

Jammet Restaurant,
Louis Jammet, Proprietor.
45 & 46 Nassau Street,
also at
Adam Court (off Grafton Street)
Dublin.

Saturday 14th

To/ Colonel McNally
o/c Curragh Command

Dear Sir.
I have just heard that an air force officer of the "Free French Forces" has been lately interned in the Curragh Camp.
I would be deeply grateful if you could let me know if it would be possible to get in touch with him, if I may send him things he might require, or perhaps see him? Would there be a possibility of having him out on parole for a few hours?
Thanking you in anticipation of a reply.
I remain yours respectfully
Louis Jammet
P.S. I believe the name of the officer is Maurice Remy

Courtesy Irish Military Archives

Canada's second official language, ordered the coq au vin, while Jack ordered vichyssoise. When the RAF internees dined at Jammet's, they dined in style.

"You see, mes amis," Remy continued, in a low voice after the waiter had left. "The IRA is our only hope." That's interesting, thought Jack. Remy, has obviously lost his mind too.

That afternoon, with Remy as their guide, the two Canadians had toured the city – Leinster House, Trinity College, St Stephen's Green on the south side of the river, and the Abbey Theater, Customs House, College Green and, finally the General Post Office, the scene of the great republican uprising of 1916, on the north. In the course of the tour Remy had given them a history of the Irish republican movement of the twentieth century, focusing on the differences between the Irish army and the Irish Republican Army.

"Surely you're not suggesting that we approach the IRA?" Jack asked. While the IRA, established as an off-shoot of the Irish Republican Brotherhood in the early part of the century, was not as active in the South during the war as they would be in the North after it, it had nonetheless made international headlines four years earlier by exploding a bomb in a fish market in Coventry.

"Yes and no, Jack. The IRA may well approach us first."

Seven had died in that blast. And anyway the RAF had nixed any external assistance after the MI9 break. Using the IRA would no doubt be nixed too. So what was the point? Just then a young woman in her early twenties with long red hair entered the restaurant, looking over at their French colleague. Remy's eyes acknowledged her and quickly returned to them.

"What was that all about?"

"Her name is May, Jack. She works for Jammet, but she also knows the right people, if you know what I mean."

"The IRA, you mean?"

"Yes, Jack but softly, si'l tu plâit. Jammet knows nothing. I do not think he would appreciate one of his staff having such contacts."

Jammet emerged from the kitchen to greet them and, just as quickly, disappeared. After a sip of cold claret, Remy leaned in towards the others. "You see, mes amis, the IRA will do anything to embarrass the government. While de V was IRA himself, he is now disliked by his former colleagues because of his interment policy." As they now knew, Remy was referring not to their internment but to the internment of 500 IRA members a mile down the road from them on another part of the Curragh's main base. "While most IRA members support neutrality, their first goal is independence for a united Ireland. They do not feel that de Valera is doing enough. So anything that embarrasses de Valera suits their purpose. And ours." The IRA internees didn't get parole. They were lucky if they got food and water.

Keefer, like Calder, was aghast.

"Those guys are terrorists!"

"Yeah. And how would our escape affect them?"

Still, getting to know the redhead setting tables across the room might be okay. She didn't look too dangerous.

"You see, mes Canadiens, if all of us escape and fifty Germans are left next door," (in fact there were fifty-eight at that stage of the war) "it would look as if de Valera planned it that way. Many believe he favours the English over the Germans, and this is not what the Irish want to hear." Remy's impeccable English and command of Irish politics was impressive. "If we all escaped, it might cause de Valera, who is a moderate, to lose the next election, or create such instability that the IRA will be able to seize power without an election. Especially if it looks like England might lose the war. And who knows, mon dieu, maybe we will. The IRA has always believed that England's difficulty is Ireland's opportunity. And in this case, England is in difficulty. So they are waiting. And one more thing." Remy paused for a final glance. "We will never be released, no matter what the diplomats say, no matter what you or the English think about the Yankees and their influence on de Valera. We are window-dressing. They need us and the Germans to protect their neutrality, their independence. Don't you see? So why not? What do we have to lose?"

Boy, thought Jack, what a story that would be.

The following day the two Canadians discovered two things. First, that another officer had arrived, a fellow-Canadian from Calgary, Flight Lieutenant Grant Fleming, who had run into foul weather west of Doonbeg off the coast of Clare the night before, and, second, that their American colleague, Wolfe, was missing.

"Do you think Ward ratted on Bud about the letter?" Keefer asked, as they reached the NCO mess. Their non-resident commanding officer, Flight Lieutenant Ward had called an "important" meeting for 11:00 hours in the NCO's eating hall, the only room in the camp large enough to accommodate them all. No one knew the details, but Keefer's guess was as good as any.

"I don't know, Bobby, but if he did I'll break his bloody neck."

"You're damn right we will."

The two Canadians seated themselves at the back by the door. The meeting, the first of its kind, was to feature the attendance of some distinguished visitors from Dublin. Ward came and sat down beside Jack, one over from Keefer. A tall man in his late twenties, with dark eyes and thick eyebrows, Ward was distrusted by most of them and not just because of the special parole arrangement allowing him to sleep off the base with his wife. He was a stickler for

Flt Lieutenant L. J. Ward.
Courtesy Bruce Girdlestone

rules, and if there was one thing that got your average RAF pilot's back up, other than a couple of ME109s on his tail, it was that.

"Sorry chaps, that we haven't had much of a chance to chat and all," Ward whispered, leaning over towards the two. "How are you getting along, Jack and Bob is it?" Ward had stuck out his hand in Keefer's direction and then just as suddenly withdrawn it. "Now, Jack," he admonished a moment later, "I've heard through the vine that you were a scribe before the war. Jolly good. But, um, you should realize that the RAF would not take too kindly to any articles published while you're here. Especially about parole, or this unfortunate Wolfe business, God forbid."

So that's it, thought Jack. For a moment, he wondered what other secrets Ward was privy to.

"The last thing we need is any controversy. I've been meaning to mention that to you for some time."

Before Jack could say anything, Ward abruptly stood up in mid-sentence and walked towards the mess entrance. The three visitors had just arrived: Sir John Maffey, the British Representative to Eire,

John Kearney, the Canadian high commissioner and Wing Commander Begg, the RAF's air attache in Dublin. Mr David Gray, the American minister to Ireland, who should have been there, had once more been called away on urgent business, this time to London.

Ward greeted them effusively and then led them quickly to the front of the room, past the thirty-odd internees who had assembled for the occasion. Following behind, as usual, were the bodyguards dressed in dark greatcoats and bowler hats. Two of the three visitors sat in chairs at the front, while the third remained standing. He was the one Ward was introducing.

"Thank you for the kind words, Flight Lieutenant Ward," said the man a moment later. "Gentlemen, I have come here today on a matter of some importance." The speaker was Sir John Maffey, British representative to Eire. He was tall and thin, in his sixties, with a distinguished mustache and a superior air. He would have been an ambassador, or a commissioner, had the English and Irish governments been able to agree on their relationship with one another, but since they couldn't he was a representative.

"Yesterday evening," Maffey continued, "Pilot Officer Wolfe of the American Eagle Squadron arrived at our RAF Belfast base claiming to have lawfully escaped. Although an investigation is ongoing, it seems likely that we will have no choice but to return him. For he did not lawfully escape, chaps. He never properly signed his parole. He'll be returned and after that, the RAF will discipline him." Maffey coughed and cleared his throat.

There was silence in the hall.

"Let me put it this way, chaps," continued the British representative. "Pilot Officer Wolfe is claiming that he has escaped by a 'ruse de guerre,' if you will. He claims that he fairly tricked the authorities by returning several times to the police hut yesterday evening, the last time on the pretense that he had forgotten something, a glove, I believe, thereby confusing the attendant, Corporal O'Flanagan, or O'Reilly, or Fitzgerald, I believe, who … well … it gets a little confusing here chaps, but basically Wolfe's slip landed up getting cancelled somehow and now Wolfe claims that he was off parole when he was finally allowed to proceed through the gate. Now that sort of thing is hardly cricket."

"Good for him," whispered Jack. They had, after all, discussed it. Many of the other internees didn't see it that way however.

"As I say, Wolfe is now relying upon this 'ruse de guerre,' claiming that he was free to escape and refusing to return to camp."

The noise increased as some abandoned their whispers, whistled their exclamations, even cursed.

"Now this conduct is hardly becoming an RAF officer, and Wing Commander Begg will have more to say about that in a moment. From my perspective it is important to recognize the Irish attitude towards the Emergency" – the diplomat paused before an audience still unfamiliar with the concept of a world war being called an Emergency – "which, as you know, poses a delicate problem for Ireland. The Irish, and in particular An Taoseach, feel that the Belligerents should adhere to the rules that the government has attached for the duration of the Emergency. If the Belligerents fail to adhere, then it is quite possible, chaps," he paused for dramatic effect, "that the benevolent side of neutrality, which you chaps have enjoyed thus far, will disappear."

Silence again. To Keefer, years later, there was something almost comical about the diplomat's terminology, de rigueur though it was at the time. It took him a while to explain it to others but the Emergency meant the war, the Taoseach meant de Valera, the belligerents meant the internees (who were growing more belligerent by the moment), and losing the benevolent side of neutrality meant losing their parole. Jack, the former reporter, leaned forward in his chair. He had a question. In the row immediately in front of him were Shaw, Welply, and Midgely, all sitting in respectful silence. On second thought, maybe it could wait.

Sir John's voice was now ringing through the mess. "Now the Irish attitude towards the Emergency has been consistent throughout – I am certain that you chaps would acknowledge as much. The actions of Pilot Officer Wolfe throw all of that into jeopardy. It is the spirit and not the letter of the parole oath which is essential to keep in mind."

"You're bloody right, sir," said Welply aloud. "No English officer would have ever done that," which drew nods of approval from most of the internees present and even scattered applause.

"Denys is right, chaps," agreed Shaw. "Wolfe had no right to breach his parole. War or no war, there are rules."

"Bang on," Sir John continued, more confident now, "I don't need to remind you that although some internees have got out of here successfully in the past, it is imperative that you do as Officer Shaw suggests, play by the rules."

"But, sir," asked Jack, deciding to push ahead – he'd been to more than a few press conferences in his time, so interrupting didn't bother him in the least – "it is our duty as members of the RAF to escape, is it not, sir? Wing Commander Begg?"

Begg spoke briefly with Maffey and Ward before answering. He hadn't met Calder, he had only heard of him – and that was enough. "It is indeed your duty to escape, Calder," Begg replied, cautiously. "We're short of airmen these days, particularly with your experience. You Canadian fellows are a great asset to us and I mean that sincerely. The difficulty we have is with the political side of things."

"What Wing Commander Begg means," Maffey interrupted, "is that any escape has to be strictly by the book. If you breach your parole," Maffey continued, driving the point home, "you'll be sent back to camp."

"But Sir, we discussed that opportunity with Wolfe." Suddenly all eyes were on Jack. "Yes sir," Jack continued undaunted. "We noticed that Captain Fitzpatrick sometimes left the police hut."

"Calder, that has nothing to do with it. The point is that it is a system of honour, and what Wolfe did was wrong."

"No sir, let me finish, please."

Oh Christ, thought Keefer.

"We felt that if one of us got out without a signed form, however that happened, sir, we wouldn't officially be on parole. It would be as if we had scaled the wire."

"Wing Commander ..."

An interruption had come from one of the two men seated at the front. The voice was that of John Kearney, the Canadian high commissioner to Eire. Kearney was forty-eight years old and a personal friend of de Valera's. He had been sitting at the front, quietly studying the mood of the group. Of the various diplomats and politicians in Dublin associated with the internees, he would turn out to be the most supportive, particularly of the two Canadians – he would offer his home in Dublin to them, his Irish wife would cook for them and introduce them to young Dublin socialites – a man who was proud to play the role of surrogate father, even offering his London flat if and when they ever got out. And he liked what he saw of the spunky reporter.

"It is somewhat premature," Kearney continued, "to suggest that

Pilot Officer Wolfe was in the wrong. Now I know that Flight Lieutenant Ward reported it" – so that's it, Keefer thought; what else did Ward tell them? – "and in all likelihood, Wolfe will be returned. But it is important that, he, Wolfe, be given the opportunity to defend himself."

"Still, gentlemen," Sir John Maffey was seeking to re-establish his position, "whatever happens to Wolfe, he has committed a most serious offence. So don't take any chances. And keep in mind that we are working hard for your release. With the Yanks on board, Churchill and this chap Roosevelt may be able to persuade de Valera to open up the ports for our boats. And then you lads will be sent back to your squadrons."

"Pardon me, sir, but haven't we heard this before?" said Covington, surprising just about everyone. "It still doesn't sound as if you're terribly interested in us escaping, sir."

"Well, Covie," Maffey said, attempting to strike a chord of familiarity. "Let me be perfectly clear. We'd prefer no escape at all, to an improper one."

Well, that was pretty clear, thought Keefer. Whose side was this guy on anyway?

"Gentlemen, I realize my words are not entirely welcome to some of you. Nevertheless as English officers you are obliged to live by the rules of parole and the other incongruities which we face in Ireland at the present time."

Maffey then proceeded to review the facts and state the British government position on what came to be known as the Wolfe affair, a position which would lead to Wolfe being returned to camp, with a formal enquiry to follow.

As the meeting continued, Keefer wasn't exactly sure what incongruities Maffey was facing. The senior diplomat had his nice house in Dublin, and in another year or two he would be posted to Ceylon, Burma, or wherever the next stop after Ireland might be for a senior English diplomat. Moreover, Maffey had just presented himself to both the pilot officer and the former reporter as yet another Englishman who referred to England as if it was the centre of the universe, and such an assumption was beginning to wear thin. It might not have been that meeting in particular, or maybe not even two days later after Wolfe had returned to camp, when the two had heard the American's side of the story and come to understand

COPY/

ROINN GNÓTHAÍ EACHTRACHA
Department of External Affairs

BAILE ÁTHA CLIATH.
Dublin.

Secret.

15 Eanar, 1942.

Rúnaí,
Roinn Cosanta.

With reference to your minute S/231 of the 30th December regarding an application from Pilot Officer R. L. Wolfe, an internee at the Curragh Camp, for permission to return to the United States on general parole, I am directed by the Minister for External Affairs to state that he is advised that it is doubtful whether international law would allow of compliance with such a request in any circumstances, and that, if it does, the obtaining of the consent of the other belligerent is a prerequisite condition. Even if it were considered expedient to seek it, there is no likelihood of the agreement of the other belligerent being obtained in the present case.

In the circumstances, I am to suggest that Pilot Officer Wolfe should be informed by the Camp Commandant that it is regretted that the request made in his letter of the 2nd December can not be complied with.

(Sgd.) J.P. Walshe.

Rúnaí.

The Irish government's formal reply to Wolfe's letter of 2 December 1941, requesting parole in the US. During the war J.P. Walsh served as Eire's secretary of state while de Valera served as both Taoseach and minister for external affairs. Courtesy Irish Military Archives

I desire to acknowledge your letter of the 25th February enclosing joint memorandum issued by the British Representatives and American Minister regarding the case of P.O. Wolfe, and also to acknowledge:-

(1) Minutes of conference in Minister's office, on 23/2/'42.

(2) Minutes of a discussion in An Taoiseach's office, on 23/2/'42.

(3) Draft copy of parole form for British Internees.

With regard to the matters to which you have particularly drawn attention, I desire to reply as follows:-

(1) Parole Forms will in future be signed in the presence of an Officer of the Camp Staff. You will appreciate that this is very difficult in view of the limited number of officers on duty. At the time of the escape there were only two officers detailed for "B" and "G" Camp duty, so that it can be readily realised that there were times when it was difficult to have an officer present in the police hut when the parole forms were being signed.

(2) The parole form duly cancelled is now being returned to the Internee when well within the compound. As a matter of fact the Internees usually run off without waiting for the cancelled parole form, others say they don't want them.

Excerpt of a memorandum from Colonel McNally to the adjutant general in Dublin, 28 February 1942, regarding Wolfe's attempted escape. During the 23 February meeting, after receiving Grey's report (next page), de Valera was asked by the American minister to reverse the RAF decision and release Wolfe. De Valera declined, releasing another "belligerent" American flyer instead, Pilot Officer Montgomery, who had landed a few weeks later. Courtesy Irish Military Archives

February 24th, 1942

Memorandum From the Desk of David Gray, American Minister to Ireland relating to the escape of Pilot Officer Roland L. Wolfe, Eagle Squadron, RAF on the 13th-14th of December, 1941.

(Excerpts) On Sunday, January 4th, 1942, after receiving permission to interview Pilot Officer Wolfe, I drove to the Curragh and presented my compliments to Colonel T. McNally, accompanied by Lt. Colonel John Reynolds, American Military Attache to this Legation. Colonel McNally received us most courteously as did Captain Fitzpatrick... He gave us every facility to see whomever we wished of the internees alone. It was evident that Colonel McNally had been placed in an extremely difficult position. He is a fine officer with great humanity and understanding, as well as force and executive capacity. Under instructions from his government he has tried to make the internees feel rather as guests than as prisoners and the internees testify to his kindness and sense of justice. They have enjoyed almost complete freedom of movement under a very liberal parole system and it is natural that Colonel McNally should feel that his good-will had been ill requited by tricks that infringed on the spirit of parole. This came out in the conversation that Colonel Reynolds and I had with him.

But it was also made plain in our subsequent talks with Flight Lieutenants Fleming and Ward that they felt themselves under obligation to their service to effect escape by every means, including ruses de guerre, short of violating parole. It was not easy for them, in view of Colonel McNally's generous attitude, but they had no alternative.

Colonel McNally assented to the theory of parole maintained by the interned officers; that is to say that the parole was given up to a certain hour and that return to the compound automatically terminated it even though such return was prior to the hour named. But Colonel McNally's version of the affair, as obtained from his sentries who checked Pilot Officer Wolfe out the last time, was that Pilot Officer Wolfe did not return as he stated for a few moments and then go out again without signing a new parole. Colonel McNally assured Colonel Reynolds and me that he was convinced that the sentries were speaking the truth and that he had promised them entire immunity from blame in the case that the truth was otherwise. It is natural and proper in the circumstances that he should support his sentries against whose trustworthiness we have no evidence. The issue is therefore drawn between the statement of Pilot Officer Wolfe and the denial of the truth of that statement by the two sentries.

After talking with Colonel McNally and Captain Fitzpatrick, Colonel Reynolds and I spent half an hour discussing the matter with Flight Lieutenant Ward, who as senior officer was in charge of the internees on the 14th of December. Flight Lieutenant Fleming, Ward' senior, who had been subsequently interned, was present at this time and acquiesced in Ward's opinion as to Wolfe. Flight Lieutenant Ward stated if he had known of Pilot Officer Wolfe's contention on the morning of the 14th, he would not have reported to Sir John Maffey in the sense that he did, because (1) it was the exploitation of an opportunity which the internees had discussed among themselves repeatedly and which they agreed was the only probable chance of effecting an escape which it was their duty to do; (2) if he had known of Pilot Officer Wolfe's contention, he would have been convinced, as he was now, that Wolfe would not have gone to Belfast unless he had in fact avoided giving parole at the time of his final exit, as otherwise he knew that he would inevitably be returned for breaking parole.

I was informed of Wolfe's absence from the compound on the day that he did not return to the internment camp. Infringement of parole is not a subject to be dealt with legalistically. Pilot Officer Wolfe's contention, however, is not a cutting of corners, but only a duping of sentries by a ruse de guerre which it was not only his right but his duty to employ, to the end of escaping and returning to combatant service against the enemy.

Flight Lieutenant Fleming, as well as Flight Lieutenant Ward, both stated that they now believed Wolfe's version of the escape as did Air Officers in the Squadron stationed in Northern Ireland. It was realised that in returning Wolfe the Air Ministry had acted on a point of honour in the light of the information then available to them and that the information now available required that action to be set aside.

Upon my return to Dublin I was convinced that it was my duty to Wolfe as an American national to sift the matter to the bottom in order that the stigma of having broken parole would not be attached to him unless the facts conclusively indicated that such was the truth. I thereupon wrote Pilot Officer Wolfe requesting him to send me a sworn statement of his version of the events of the night of December 13-14...With the direct testimony of two American officers serving in the Royal Air Force explicitly alleging that Pilot Officer Wolfe returned a third time within the compound and went out without signing a new parole, I feel called upon, at the least, to take such action as well bring the available supporting evidence before Pilot Officer Wolfe's superiors and to make such report to the Department of State as will tend to safeguard his honor as an officer and gentleman.

Signed, David Gray, American Minister of Ireland

Courtesy Irish Military Archives

Ward's role in all of it, but eventually each would come to question the whole manner in which the English had treated Wolfe. Their rigid sense of propriety and custom had disallowed any admission of responsibility, the enquiry that followed had been a charade, forcing Wolfe to write a report, using it against him, and then ignoring what the American minister concluded; and also, the petty matter of their refusal to recognize that, given all the bad weather that fall, the allied internment camp on the Curragh now had officers and NCOs from Canada, the United States, Poland, France, New Zealand, Australia, and even Northern Ireland. In fact, the non-Brits outnumbered the Brits and, after Pearl Harbor, it was no longer just their war. And the irony of the whole thing was that if it hadn't been for the English, the Irish wouldn't have interned any of them. Or so they believed.

It was after Wolfe's return that the celebration finally ended, that the reality finally sank in. By then their hopes for immediate release had disappeared. Most sided with Wolfe eventually, except for a handful of the English officers.

As for Keefer, he spent the balance of the year on parole, coping with the disappointment as best he could – on the camp links, in Newbridge and Naas, and in Dublin for another dinner at Jammet's, where he and Jack had the pleasure of meeting May, their future IRA conduit, leaving them both wondering, in the aftermath of the Wolfe affair, whether Remy might not be right after all. Flight Lieutenant Grant Fleming and Sub-Lieutenant Bruce Girdlestone, who shared the hut next to theirs, often accompanied them and the four became good friends.

Of course Susan Freeman was on his mind through much of this period as well, especially in the run-up to the Harrier's Ball, the big New Year's dance at Osberstown. He knew he had left her in the lurch and was feeling a little guilty. Maybe now that they would be around for a while, he could make amends. The Harrier's Ball at Osberstown was always the highlight of the hunting season in Kildare. While the Curragh Grandstand hosted the thoroughbreds, Osberstown hosted the hunts and point-to-points, providing welcome relief for the Planters as they galloped about from estate to estate in their jhodpurs. Keefer found himself drawn to these events, almost against his better judgement. So, after once more refusing to get dressed up as one should have, he and Jack set off on their bicycles to join in the hunt and then ring in the New Year. The hunt, with Susan at his side, was even more pleasant than the last time.

After the hunt, he and Jack continued their good luck, finding themselves at a table with Susan and a friend of hers, Ann Mitchell, of Ballymore Eustache. Ann was the daughter of Major John Mitchell, an ardent supporter, and the president of the Kildare Hunting Club that year. She was also a member of England's Women's Auxiliary Air Force and had served briefly in England before returning home for the holidays.

"So you're the reporter that everyone talks about."

Jack, never at a loss for words, suddenly found himself struck silent in her presence.

As for Keefer, he soon found himself out on Osberstown's spring maple-parquet floor, dancing up a storm with Susan, resplendent in a long white lace gown and red elbow-length dinner gloves. New Year's Eve was always a special night for him, a night when people celebrated without fear that their expressions of affection might come back to haunt them. A romantic night.

"So, Bobby, what are your plans after the war?"

"Maybe I'll be right here," he said uncertainly, as her left hand tightened around his collar and her body pressed against him. Initially, he had tried to duck the question, and then he mumbled his half-hearted reply. Now he was hoping she didn't get the wrong impression.

After another few waltzes, she moved closer and asked him again. He struggled to say the right thing. He hardly knew the woman; nor did he know how long he was going to be stuck there. What was he supposed to say?

"You don't have to go back to Canada, do you?" she said, not believing him for a moment. "You don't have a girl friend do you? I think you would have already told me that. You're not one of these RAF lads who are are married, are you?"

"Certainly not," replied Keefer. How could the woman think such a thing!

"Then what's the problem? Don't you find Ireland a beautiful country, a wonderful place to live, a wonderful home for your children?" He swallowed hard. "You will have children one day, won't you?" No response. "Don't you find the more time you spend here, the less you want to go back? To Canada, I mean."

Susan had considered joining the WAAFs, like her friend Ann Mitchell, only she thought she'd miss her home too much. "Ireland's a beautiful country you know, Bobby, if you ever get the opportunity to visit the various counties, not just the Pale, but the west and the north. You do like golf, don't you?" She squeezed him by the neck, a little too hard.

"Um, yes," he replied flustered. He couldn't stand it anymore. "Would you excuse me please, Susan, for a moment?" Such a polite young man.

As he stumbed downstairs into Osberstown's cavernous basement, Keefer expected to find Jack. Instead, in the cellar bar at the foot of the stairs, he met his first German, a Luftwaffe oberlieutnent who looked very much like he did – six feet tall, strong build, medium brown hair and rounded shoulders, dressed in a grey tweed jacket and blue cardigan turtleneck. His first reaction was fear. His second was to turn around and leave, but he kept walking towards the bar anyway, thinking that anything else might be interpreted as a sign of weakness. The German must have felt the same way for he, too, reacted with a start, but then continued, determined not to be put off by the enemy's presence.

There was one other person in the bar, the tall blond he had met on his last visit; Josephine, the bartender. After a tense moment Josephine greeted each in turn, first the German in his native language and then Keefer, whom she recognized from their previous meeting. She gestured for them to sit down on the two stools in front of her. The German sat, the Canadian remained standing. Josephine then set about the business of pouring pints. For the next few minutes they remained there, the German drunk, or apparently drunk, and Keefer frozen in his tracks, reluctant to go back upstairs quite yet but unwilling to sit down either. Neither smiled or said a word. Once the pints had settled, Josephine placed them down on the bar, nodding at each in turn, as if they could now speak. The German spoke first.

"How long have you been in Ireland?" he asked in good, almost perfect, English. "How long has the RAF kept you here?" he repeated, in what outsiders might have concluded was another innocent attempt at conversation.

"It's not the RAF keeping us here, it's the Irish," Keefer replied.

They sat silent for another minute.

"Doesn't it strike you as odd that we have nothing to say to one another?" the German said. "Well, doesn't it? We are enemies you and I, and yet here we are, alone in a bar in a neutral country, and in one or two hours of discussion we could probably end the war. Do you realize that?" The German seemed drunker than Keefer first thought, slurring his words and gazing past him. "So let's get started, shall we? My name is Conrad Neymeyr, I am an oberlieutnent. It is a pleasure to meet you ..." The German had stuck his hand in Keefer's direction, waiting, before retracting it in disgust. "You see, you have no courage. Josephine, bring my friend a bottle of Schnapps. We will see what we can do about this problem of no courage."

"No we won't, sir," said Keefer, pushing the pint away. He felt a different kind of fear now, not of being attacked or insulted but of being found out, of being seen in such a compromising position. So he left. Maybe he was safer on the dance floor. Climbing the stairs, he bumped into Covington.

"What were doing talking to him, Bobby? Neymeyr? He's a bloody parachutist, you know that?"

"Well no, but ..."

"The chap in the blue turtleneck? You talked to him. Christ, Bobby, I saw it. He's a spy. He always pretends to be drunk – that's his job. He knows more about the Irish than the bloody Irish do. In fact, rumour has it that he drank a whole team of Irish investigators under the table. Honestly, lad, be a little more careful next time."

It took some time for Keefer to recover from Covington's warning and return to the ballroom to face the music. When he did, he spotted Susan alone in the corner, with her mother. He smiled, and she smiled back, and he realized that if he wasn't interested in her, he wouldn't be doing that, for there were lots of other women around. So he asked her to dance.

Susan, for her part, didn't mind. She wondered whether he had suddenly shied away because of something she might have said. Her mother had always encouraged her to be direct. Was that the problem? For in truth, while she wouldn't have minded staying in Ireland, she wouldn't have minded leaving either.

So it wasn't a surprise to either the pilot officer or his audience years later that the evening ended, among all this uncertainty, on a

2 December 1940
Confidential
C.S.O. i/c
G2 Branch

RE: German Aircraft - Inishvickillane

Sir,

I have the honour to report the following:

As instructed by you, I went to Cork with Commandant Delamere on 29.11.40, arriving at Collins Barracks at 19:10 hours. On arrival we went direct to the Duty Room, Officers Mess, where Major O'Connell, Captain O'Donoghue and 2/Lieutenant O'Sullivan were with Oberlieutnant Neymeyr of the German air force. On being presented "as from Army headquarters" I told him that we had been sent to report on his condition and that of this crew to ensure that they lacked nothing, and to obtain a complete list of their personal data to be transmitted through External Affairs to Herr Heinkell, the German Minister, in order that their relatives be informed as soon as possible of their safety.

Oblt. Neymeyr was 24 years of age, of good physique and education, had joined the German army on leaving high school in 1936, and had two years flying experience probably in connection with the navy as he seemed acquainted with naval matters and often referred to the navy. He had the Iron Cross conferred on 22.11.40. He was very friendly disposed and a cordial conversation developed which later became general and which was very astutely guided by Commdt. Delamere into a friendly argument of flying matters. He was naturally not asked any direct questions on matters which could be considered confidential, as at an early stage, when a question as to the type of engine he had arose, he stated: "You will appreciate my position; I trust you completely, but we are warned to give no information whatever lest it might fall into the hands of our enemies." He then stated that he was not sure, as he was not skilled in engineering matters but he thought his engine was a Benz.

Contrary to expectation, the consumption of seven glasses of Irish whiskey seemed to have no noticeable effect other than diuretic...

Regarding the events which led up to his mishap he was not disposed to give any information and was therefore not pressed on this point. He stated it was due to engine trouble and that Commdt. Delamere would understand that usually when there is trouble with an engine, fifty other things also go wrong... Oblt. Neymeyr possesses in my opinion more than the average Continental's knowledge of Ireland - he knew of the partition situation, he knew that the Irish people are predominantly Catholic and he knew of the Irish dislike of the English. He enquired if the Irish were interested in the racial problem, and if we had any trouble with the Jewish problem...

Your obedient servant,

Sargeant Jos Healy, G2

Report of Neymeyr's landing. Note Sgt Healy's reference to the diuretic, fourth paragraph. Reports of this nature were often used by Keefer to convince his audience that he hadn't made up certain parts of his story, referring here to both the fact that Neymeyr was likely a spy, as well as his own detention and interrogation at the hands of the Local Defence Force in county Clare. Courtesy Irish Military Archives.

high note, he and his Anglo-Irish-Jewish girlfriend retiring to the Commowealth Room upstairs, with its bay windows overlooking the English garden. After Jimmy Dunny and his celtic-swing-fusion band had packed up and left, and after her mother had ordered Patrick to remain behind, to drive him and Jack back to camp. It was a huge room with peach wallpaper, he remembered, oak wainscotting, and a floral sofa in the corner. The commonwealth room, where they kissed for the first time, where war seemed little more than a memory.

The oak shutters rattle with the wind and the lamp burns dimly in the corner. The rain picks up, thundering against the tin roof above his head. The screen door slaps open and closed, telling all who care to know that he is still there.

He is lost, sinking, grabbing for the Mae West as the man in the rubber raft wearing the blue turtleneck sweater watches, smiling.

He is back in his room. Suddenly Virtue bursts in. "Diaper's escaped, sir, Diaper's escaped. He's tunnelled under the main gate, he's made it, sir. We'll all make it, come, hurry."

"Piss off, Virtue," he answers, from his bed, book in hand, eyes glued to the page, "I ain't going nowhere." The shadows of Dalton and Brown appear at the door. He sits up. "What do you want with me," he snaps. "You're just going to get yourselves killed, you know that?"

"You got us here, sir," replies Virtue, frightened, "you get us out."

Brownie grabs him by the shirt. "Why aren't you helping us, you bloody bastard?"

But he doesn't care. He holds on to the mattress until they leave, with only their stares to remember. A few minutes later Jack appears at the door, smirking. He begins to cry as Jack turns away in disgust.

The next morning he wakes up and goes directly to the police hut. The gate is open and all the guards have left. The internees have left too, all except the Baron who sits in Fitzpatrick's seat smoking his stupid cigarette. The Baron waives him through but instead he reaches in and grabs a parole slip from beneath the counter, signing it as many times as he can – just to be on the safe side. The Baron then reaches out for his naked body (for inexplicably, he'd forgotten to dress) but he runs away at full speed, out the main gate and then down the road, passing Germans along the way, eight abreast, laughing at him, challenging his manhood, pointing at it.

He is now on the fifty-yard line, buried beneath a wall of Western Mustangs. It is 1937 and he is in his second year of college. He is groping around for the ball. His team has lost seven games in a row, but he is undeterred. He doesn't mind losing game after game, as long as he gives it his best shot. No one could call him a shirker, then.

Suddenly he is back in Ireland. Climbing the winding drive beneath a row of sprawling oaks, he sculks from trunk to trunk. When he reaches the top a man greets him by the carriage house and commands him inside. "Sit down Pilot Officer Keefer," the man says. He sits cross-legged on the floor. Three other men enter and stand over him, two in RAF *uniform, and the third, in plainclothes, shorter and fatter than the others with a cigar in his left hand and a revolver in his right.*

"You have turned your back on your country, Pilot Officer Keefer," says the short, fat one, raising the gun and placing it to his temple. Cocking the trigger.

The pilot officer shudders from the cold steel.

"You're a shirker, Keefer," he says, slowly removing the cigar from his fat mouth.

Keefer once more begins to cry.

The man squeezes the trigger.

6

THAT LET-DOWN FEELING

“Wake up Bobby, Bud’s in trouble. Jesus Christ, Keefer! Wake up!”

Jack left his friend on the bed and turned to the window. From there he watched and listened as Wolfe and the three Englishmen hurled insults at each other, something which had started two hours earlier at Osberstown House soon after Keefer and his navigator had left. In a drunken stupor Ward had told Shaw about the letter, as Wolfe was sitting in the corner talking to Ward’s wife, Vi Lawlor, aged eighteen and pretty as a picture.

They had continued arguing on the way home. The dispute would be settled, they decided, on the links, by Hanley’s Cross, otherwise known as Donnelly’s Hollow, where a famous Irish boxer once thrashed an Englishman within an inch of his life or so the legend grows. The man’s arm was bottled in vinegar out of respect for the fighting Irish spirit and can still be found in a jar in a nearby pub. This time, however, it would be an Englishman teaching an American a lesson about honour, and what it takes to be an officer in His Majesty’s Royal Air Force.

They had settled nothing, so they had continued the duel in camp.

And now, as Jack watched, the American was backing away, his eyes darting from one adversary to the other.

“You think that will do you any good, you coward?” said Shaw, as Wolfe removed his scarf and outer-coat, winding them like propellers above his head before flinging them dramatically to the ground. Within seconds Wolfe’s shirtsleeves were soaked in the driving rain.

"Come on you bloody shirker, let's settle this once and for all," said Shaw, now circling the American with fists cocked, still dressed in black tie and full eveningwear.

"You've got a deal, you wart off a horse's ass."

Just then Midgely crept in from behind Wolfe and pinned the American's arms at the elbows. Jack jumped through the open window, seven feet to the ground.

"Leave him alone, you coward," he shouted at Midgely, running towards the circle. Shaw was punching the American as Jack arrived, driving the Englishman to the ground with a shoulder block. Quickly Shaw was back on his feet, but Wolfe was free now too, poised to meet his aggressor. They came together in a drunken fury. Shaw began with a left to Wolfe's head, just as the American struck the Englishman with a hard punch to the stomach. Each fell back from the force of the other's blow.

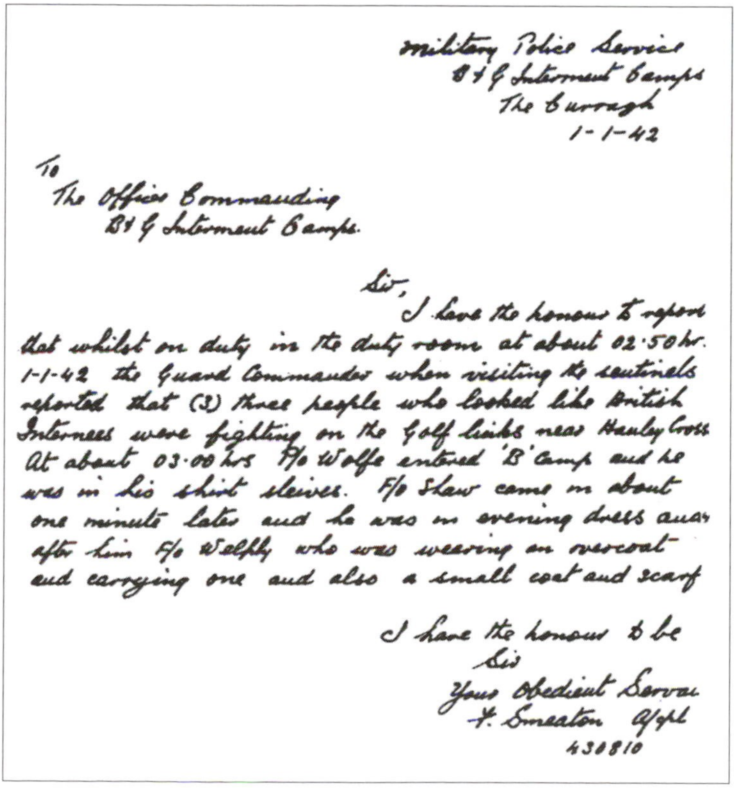

Military Police Service
B & G Internment Camps
The Curragh
1-1-42

To
The Officer Commanding
B & G Internment Camps.

Sir,
I have the honour to report that whilst on duty in the duty room at about 02·50 hr. 1-1-42 the Guard Commander when visiting the sentinels reported that (3) three people who looked like British Internees were fighting on the Golf links near Hanley Cross At about 03·00 hrs F/o Wolfe entered 'B' Camp and he was in his shirt sleeves. F/o Shaw came in about one minute later and he was in evening dress and after him F/o Welply who was wearing an overcoat and carrying one and also a small coat and scarf

I have the honour to be
Sir
Your Obedient Servant
F. Smeaton Cpl
430810

Cpl Smeaton's report of the duel. Courtesy Irish Military Archives

Donnolly's Hollow/Hanley Cross where an Irishman once thrashed an Englishman to within an inch of his life, or so the legend grows. Fifteenth tee is in background. Courtesy William Gibson

"You'll see more of that my good man, before I'm done with you," said Shaw a second later, lunging at Wolfe, using his greater height to advantage, taking three or four quick shots at Wolfe's face, pinning his target with a headlock, connecting with his right hand. He continued, blow after blow, not in the least assuaged by the blood now dripping from the American's face.

Wolfe grabbed Shaw by the waist and flipped him over. The Englishman fell heavily, his eveningwear splashing in the mud, the white collar of his tabs stained red with the American's blood. In a surge of strength, Wolfe climbed astride the Englishman, kneeling over him, pushing the side of his face into the mud, pounding him with both fists. As Jack stood back, prepared to intercede, Grant Fleming, their new commanding officer, scrambled out of the hut next to Jack's and ran towards the melee.

"What in hell is going on?" he shouted, pulling Wolfe off Shaw and throwing him to the ground. Shaw went back at the American, but Fleming kicked the Englishman squarely in the groin, doubling him over in pain. A second later Wolfe went back at Shaw, pulling the outer coat over his head, throwing haymakers as the Englishman remained doubled over, groaning helplessly.

"Stop it," Fleming commanded once more, grabbing Wolfe by the hair and yanking him off the Englishman. "Now what in the hell is going on!"

Neither spoke.

"Well, Wolfe?"

"I think you'll have to ask Flying Officer Shaw that, sir," replied the American, crawling on all fours to the hut behind them, then climbing to his feet. By then Keefer, Girdlestone, Remy, Karniewski, and a dozen NCOs had arrived.

"Well, Shaw?"

The Englishman stood up, seeking to regain his dignity. His tabs had been ripped in the struggle, his footwear and trousers were covered in mud, and his cummerbund had slipped up beneath his arm pits – a dreadful sight indeed.

"Wolfe, sir, is a shirker," he announced formally, before taking another run at the American – only to be stopped, this time with a sharp cuff to the back of the head.

"Next time it's the face," threatened the flight lieutenant.

"I say that, *sir*," said Shaw, unrepentantly wiping the mud off the lapels of his dinner jacket, "on account of the letter. Yes, sir, apparently Wolfe told our good friend de Valera that he was 'strictly for hire.' In my book, that is reason enough for a sound thrashing."

"A mercenary, is it," said Welply, moving towards the American.

"How dare you." said Midgely, closing the circle.

"Stay right there, you three," said Fleming, menacingly. "I've known about that letter for a while and in my book, Pilot Officer Wolfe is no shirker. Did you others hear that? He's no shirker." Fleming remained standing in the centre, among the dozen or more men. "Wolfe has every right to be frustrated and it's too bad you all weren't a little more frustrated. And I'd gladly discuss that with any of you, one on one."

No one moved.

"I'm sure Wolfe wouldn't write the same letter today," Fleming added quickly, "and if he did, I'd be the first one to beat the piss out of him." Wolfe looked away, avoiding Fleming's stare. "Things have changed – you got that? All of you?"

Fleming had turned a full circle to ensure that each of the

internees, including the NCOs, had heard. "This isn't just England's war any more, it's all of ours. And it's time we put our heads together. We've got to get out of here – and we're going to."

The rain fell hard in the shadow of the compound's searchlight. Their new leader had been watching his colleagues since his arrival three weeks earlier, eating with them in the mess, drinking with them in the officers' bar, and, unlike his predecessor, actually sleeping with them in camp. He had been looking for the right opportunity. And this was it. It was one thing to make the best of their situation, he had decided. It was quite another to enjoy it.

Fortunately, the adjustment for the two Canadians wasn't that difficult. In Keefer's case, he was primed to get back to the war as soon as he could; as a volunteer, he would not have been there had he not felt that way. Not that he would have minded spending the next few years on the Emerald Isle under the firm grasp of a woman like Susan Freeman. No, that mightn't have been too taxing. Only that he might have found living with himself after that difficult, knowing that there was a war going on. Susan Freeman could come later, if he so chose. And when it came to choosing, of course, he could certainly do worse, since she was suitably endowed, physically, intellectually, and financially, for the harrowing task of raising his children, of perpetuating his family name. It was more a matter of timing. With Flight Lieutenant Ward, there had never been an issue. But with Fleming, the time had clearly come to stand up, to be counted.

Jack, like Keefer, was also getting bored. Even the distraction of waiting to hear back from his editor at Canadian Press, thinking that his harmless little account of their landing in Clare must have hit a snag, provided little relief. If Flight Lieutenant Grant Fleming could get everyone pulling in the same direction, then so much the better. He, too, would far prefer to be back in the war, writing copy from where it mattered most, looking down the chute, than stuck in this God-forsaken place.

If it wasn't obvious to the former reporter, it should have been obvious to the rest of them: the hero of the story had just arrived.

It should have surprised no one, really. Flight Lieutenant Grant Fleming, from Calgary, Alberta, had been a war hero long before he arrived in Ireland, one of just a handful of Canadians who had already received their DFCs. He had earned his in a daring daylight raid on the *Bismarck* in the summer of 1940, when he had flown a four-engine Sunderland flyingboat low over the German battleship's deck, providing his passenger, an admiral of the Royal Navy, a first-hand look. This was followed by a harrowing thirty-two hours ferrying a twin-engine Catalina flyingboat across the Atlantic, normally a sixteen hour crossing, but not after losing both your ailerons.

Nor was Fleming's arrival in Ireland any less spectacular. After ditching off the coast of Clare, he had survived a two-hour, three-mile swim in the frigid waters of the North Atlantic. The only other member of his crew of eight who survived was Jimmy Masterton, an Olympic swimmer from Perth. All that Fleming remembered was waking up on shore and discovering that his watch had been stolen.

So when Grant Fleming came to camp, he was in no mood to golf, bet on the horses, frolic with the daughters of the Protestant gentry, or merely think about escaping.

In the next few weeks, the four of them – Keefer, Calder, Girdlestone, and Fleming – took parole together at every opportunity and talked about escaping. While there had been a lot of whining among both volunteers and permanent officers about the armed guards, the amount of wire, and the Wolfe affair, there hadn't been a lot of thought given to the reality of their situation. Sure the guards were armed, but would they fire? If so, would they use blanks? And even if they did fire live ammo, they were clearly understaffed. So could the internees withstand an attack nonetheless? With only a mixture of police and army sentries to confront them, the internees likely outnumbered their gaolers on any given night. The key, therefore, would be a plan that involved everyone.

And if someone got shot in the process, hopefully it would be an NCO.

Jack's job, as the reporter among them, was to research previous attempts and report back to the group. His sources would be his fellow internees, the diplomats, the gentry, and whatever information he could glean from the army, including Colonel McNally. Given that no one had tried to escape more than Covington, and succeeded less, Jack began with him.

It was 7 January, when Jack first cornered the Englishman in the officers' bar, doing his monthly accounts at his customary post by the window.

Crash of British Sunderland Flying
Boat at Carrowmore, Co.Clare,
on 3/12/1941.

At about 18.30 hrs. on 3rd December a British Sunderland flying boat crashed on the sea about 2½ miles from land at Carrowmore near Doonbeg, Co. Clare. Two of the crew of eleven made their way to the shore but the remaining nine occupants were drowned. The aircraft was later washed ashore on the strand in a wrecked condition.

The particulars of the survivors are as follows:-

(a) 40380 Flight Lieut. James Brandt Fleming D.F.C.
Age 21 years: Religion Presbyterian.
Next of Kin: Father, address: c/o Dr. S.C.W. Morris,
Souchen Buildings,
Calgary, Canada.

(b) 911625 Sergt. James Cannell Masterson,
Age 21 years: Religion Church of England.
Next of kin: Father, William.
Address: Swanton, Abbot, Norfolk, England.

Both survivors were taken into military custody on arrival at the shore by a military patrol and after being held overnight at Doonbeg were transferred to Mallow Military Hospital on the 4th December. Lieut. Fleming and Sergeant Masterson were suffering from shock as a result of their immersion and Masterson had in addition abrasions on both legs. Both are still in Mallow Hospital on this date (9/12/41).

The bodies of four other members of the crew were washed ashore near the scene of the crash and were buried locally with military honours.

Covie's first attempt, Jack learned, had occurred four days after his arrival, on Christmas night, 1940. Already fed-up with things, the bar officer decided to test the turnkeys' mettle by brazenly walking past the guard in the inner compound (this was before the police hut was built) and then out towards the wire. It took Covington only a matter of seconds to cross the wire – several layers of coils had been added since then, however – and soon he found himself racing towards the main gate. Although he'd thought he might be shot, Covie recalled only a sharp whistle before being tackled by soldiers on the road.

Jack assumed, therefore, that the question of whether the guards had authority to shoot them was still very much an open one.

Next came the 21 January break, one month later. This time five of the six internees at that time had escaped after one of the English

Ref. No. G2/X/0684.

G.2 Branch,
Department of Defence,
Parkgate,
Dublin.
4th June, 1942.

General Officer Commanding, 1st Division,
General Officer Commanding, 2nd Division.
Officer Commanding, Southern Command, (G.2)
Officer Commanding, Western Command, (G.2)
Officer Commanding, Curragh Command, (G.2)
Officer Commanding, Eastern Command, (G.2)
Provost Marshal, Department of Defence.

Army Order 43/1941 - Crashes of belligerent aircraft.

Following a recent crash some members of our forces were courtmartialled for appropriating articles of personal property belonging to the victims of the accident. The articles were a wrist watch, pocketbooks and money. We have reason to believe that there have been similar occurrences after previous crashes where gold rings and small arms were appropriated by military personnel and not surrendered subsequently.

This looting is in the highest degree discreditable to the Army, and the Chief of Staff is anxious that every effort must be made to stamp it out by all possible means. Officers in charge of military parties on the scene of a crash must exercise the strictest control over the personnel. Where there is reason to believe that any member of the party has been appropriating property he must be closely searched on the spot and the whole party should be searched on return to barracks.

Whenever possible an officer of the Provost Marshal's Branch should attend at the scene off the crash in order to keep the military personnel under observation and to assist at any searches that may be regarded as necessary.

The officer in charge of the military party on arrival at the scene should obtain a list of the property already recovered by the Garda and, in conjunction with the Garda, exercise strict control over the activities of civilians and sightseers who may be present.

[signature] /COLONEL.
(DAN BRYAN)
C.S.O. G.2 BRANCH.

The theft of Fleming's watch and two other events like it prompted the above memo. Given the transportation difficulties in Ireland at the time, the poverty, and the curiosity with which each crash was met, it is not surprising that a handful of such incidents took place. Courtesy Irish Military Archives.

sergeants tossed an iron bar onto the electric wire, shorting the searchlight's circuit. All four were nabbed before reaching Belfast, however, including Covington, who suffered the indignity of being dragged off the bus at Rathcoole.

A month later Covie was at it again, fashioning a rope ladder from bits of his parachute that he had been allowed to retain as a keepsake. Unfortunately the makeshift ladder was confiscated before he had had the chance to use it. Then on 20 March, on his fourth try, Covington got out briefly on a bicycle after duping the sentry into believing he had given his parole when he hadn't (not unlike Wolfe), only to be picked up in Newbridge. He and Ward had had it out, just as Wolfe and Ward would several months later. Finally, he had escaped during the MI9 attempt that July, his fifth stab, only he had made it no further than McCabe's pub in Newbridge, after making the mistake of sharing a few pints with an off-duty garda. "How was I to know," Covie explained defensively in the face of a particularly incisive question from the former reporter. "I assumed he was one of Nora's cousins," referring to his girl friend, Nora McMahon. But, as Jack noted, Covie had been rather quiet the past six months.

"You're right, Jack, we've got to do something, damn it." The interview in the bar had gone on for some three hours at this point. "And I'll bet you a fiver that they would never shoot us. Never!" Of course, Covington had made that bet before. "There is no way in God's sacred green Irish earth that they would dare, Jack. Could you imagine His Majesty's reaction if even one of his subjects was made the sacrificial lamb of these Irish hordes, these ..."

"Covie, nobody would give a fuck," slurred Jack.

"I think not, Jack old boy. If one of us were to be shot, the repercussions would be immediate. Indeed, it might even trigger an attempt by Winnie to reclaim the Irish ports."

"Yeah, right, Covie, for thirty flyers."

"Maybe, maybe not. But one of us must try!"

"Alright, Covie. To the Dash! We need those bases, old boy. Get out there and show us the way!"

Covington was moved to silence. It would be an exaggeration to say that his eyes welled or that his chest filled in anticipation of the challenge, but clearly the moment had arrived to show Jack, and indeed Fleming a thing or two about courage. If he waited until

"D" Company,
18th Infantry Battalion.

To/
The Adjutant,
<u>18th Infantry Battalion.</u>

<u>21st/I/'41.</u>

Sir,

I have the honour to submit to you a report about the arrest of a Flight Lieutenant of the R.A.F., by me on this morning (21st/I/'41), at approx. 08.45 hours, near the village of Rathcoole, Co. Dublin.

While on my tour of duty as Sergeant i/c. Stand-To Party, I halted a Bus for inspection of passengers. In endeavouring to do so, I was obstructed by the bus Conductor, i.e. he was rather abusive in his language and described my action in halting and inspecting the bus as a "bloody racket". After a brief exchange of remarks I boarded the bus despite the attitude of the Conductor, and the second man to draw my attention proved to be the prisoner (the Flight Lieutenant referred to above).

The Bus was the property of the I.O.C.

I have the honour to be, Sir,
Your Obedient Servant,

(Signed) E. WHELAN, Sergeant.

TQ/

<u>Copy To:- G.2., Eastern Command, for information.</u>

While Covington, whose name does not appear in the above report, claimed not to remember the incident on the bus, the fact remains that he was the only flight lieutenant in camp at the time – quite apart from the report's ring of truth. Courtesy Irish Military Archives.

Memorandum to file: 27.6.41, regarding Pilot Officer Covington's arrest (excerpt)

At 09:00 hrs on 26th instant, Flight Lieutenant Covington was seen by a Military Policeman who was in civilian clothing entering McCabe's Public House, Newbridge. The P.A. followed Covington into the shop and was asked by Covington to join him in a drink…After the P.A. secured an escort for Covington's return to camp, Covington told the P.A. that it was his intention when entering the shop to have a drink, and then procure a car in order to return to camp… Covington, on his arrest, was very lame, complaining of being thirsty and was obviously distressed.

Again, only perfunctory denial. Courtesy Irish Military Archives.

morning, the opportunity would be gone – the only memory being yet another hangover.

"Very well, Jack. You lure the policeman away from the hut – tell them you'd like to post a wire. I'll make a run for it to the painter's ladder by the shed. I'll grab that and be off. What do you say? Jack?"

"Sleep it off, Covie, I was just joking. Tomorrow, Buddy, tomorrow."

The bar officer looked disappointed. "Jack?"

"Bang on, old boy," replied Jack, changing his mind again. "No time like the present."

"Well said," replied Covington, swallowing hard and heading for the door. "If their orders have changed, then most assuredly they will aim low."

"Most assuredly."

"What harm will come of it?"

"None, Covie, none."

"So, what do you say old boy?

"Count me in," Jack replied, stumbling in behind.

Just then Fleming appeared.

"It's no good, sir, you can't talk us out of it," declared Covington, stumbling past the bemused flight lieutenant and heading out into the compound. "God save the king," said Covington.

"God save the king," replied Fleming, turning to watch the two men stumble out the door.

"Good evening, officer," said Jack politely as he reached the police hut a moment later.

"Good evening, Officer Calder," replied the policeman, who was on duty that night. (Fitzpatrick seldom worked nights.) "A little late for parole isn't it, lad?"

"Yes sir … Ah, no sir … Ah, I'd like to send a wire home to my mother, sir."

"Why so polite, Calder? My goodness, you must be up to something."

Nevertheless, the gardai shrugged and then unlocked the door to the outer compound. This served as the signal for the escape to begin.

It was not a sophisticated plan.

With a glance in Fleming's direction, Covington began his dash to the first wire. Arriving a moment later, he began his ascent, cursing and swearing. As Calder distracted the gardai, Corporal Smeaton, one of the sentries at the main gate, suddenly appeared. "Is everything all right?" the corporal asked, to which Jack replied that everything was fine. Just as Covington reached the top of the coils, however, a second guard, the pugnacious Corporal Kelly appeared. Second in command to Captain Fitzpatrick, Corporal Kelly was a man none of them liked. Fluent in German, the product of a German Jesuit school prior to the war, he was decidedly pro-German. And anti-them. In fact, he was now waiting for Covington on the other side of the first set of coils, smacking his fist into his palm.

Another policeman, who shared sentry duties with Cpl Smeaton, appeared alongside. The policeman drew his weapon and, when Covington saw it, he thrust his hands in the air over his head and began shouting.

"Don't shoot, Don't shoot!"

"Go ahead, buddy, they can't shoot, those are their orders," Jack shouted back from his post at the police hut. He had, after all, researched the issue.

Emboldened by Jack's words, Covington jumped from the top of the wire and landed at the feet of the startled policeman. He pushed the man aside and dashed across the outer compound towards the painter's ladder. He then grabbed the ladder, fumbling it into position for his final assault.

Much to everyone's surprise, especially Jack's, the policeman fired his weapon, an Ensign which he had leveled at Covington's head. Covington froze in his tracks, then slumped to the ground. The NCO's and officers of the RAF camp scrambled outside. Who had fired the shot? A soldier or a policeman? Had it hit anybody? Maybe a German or one of them, God forbid? Noises could also be heard next door. Three or four German internees had climbed onto the roof of the hut closest to the adjoining wall, peering into the enemy's camp. They were cheering.

"No sweat, buddy, they're shooting blanks," Jack yelled a moment later, as Covington picked himself up off the ground, checking for bullet holes.

Fleming smiled. He had his answer.

The next item on Jack's agenda was to interview the various mem-

Command Headquarters,
Curragh Command.

8th January, 1942.

Adjutant General,
Department of Defence,
Parkgate, DUBLIN.

SUBJECT:- Attempted Escape - British Internee, Pilot Officer Covington, 7/1/1942.

Sir,

The above named Internee made an attempt to escape at 17.50 hours on the 7th instant. He was apprehended again seven minutes after that hour. I am enclosing herewith reports from the personnel concerned.

Pilot Officers Covington and Calder were proceeding on parole. At this stage neither of them actually had signed parole form but Calder had approached one of the policemen on duty in connection with the dispatch of a wire. This was obviously part of a prearranged plan in order to allow Covington to make his dash across the wires while at least one of the police were engaged. Covington, as can be seen from the reports, climbed up the wires at the gate. He was covered at all stages by the policeman and actually when he came to the ground on the outside of the wires he had surrendered to the policeman and had his hands up and was pleading not to be shot. Calder from the inside shouted to him "go ahead buddy, they can't shoot you, their orders are not to shoot you". With this Covington dashed off. I may mention that at this stage the Guard Corporal was posting his guard and Calder further tried to confuse the issue by shouting to the Guard Corporal "not to bother that everything was all right". Covington made for a painter's ladder which was outside the compound, left at that particular spot to be under proper supervision by the police on duty. He dashed on to the protective wires and threw the ladder against these and proceeded across them. In the meantime the N.C.O. i/c. of the guard had fired a few shots in the air and this drew the attention of some of the members of the Staff who happened to be on their way home at the time. These men with a few other soldiers succeeded in capturing Covington and bringing him back. At the same time I may add that the military police had their bicycles and were in touch with Covington all the time so that his escape was practically impossible.

Colonel McNally's official report of the Covington-Calder incident, 7 January 1942. As the reader will note, the dialogue is not wholly a figment of the pilot officer, or his son's, imagination. Also, this would not be the last time that the colonel would write the adjutant general in Dublin requesting separation of the two camps, as referred to in the second to last paragraph, opposite. Courtesy Irish Military Archives.

The main point now at issue is the fact that the British Internees have definitely and for all time established that our police and sentries cannot shoot them when attempting to escape. This needless to say has placed the Internment Camp Staff in an impossible position. We must now face the situation and answer the question whether we can shoot the escaping prisoners or not. If not it would be advisable to have the police guard unarmed but equipped with batons and definite authority given to use these batons vigorously in the case of an attempted escape. In addition and in view of the fact that the men would not be armed additional personnel has been demanded by the Governor of the Camp and I am in entire sympathy with this demand. It will mean at least an increase of 12 police to man four extra posts inside the compound. I consider that two additional Officers are necessary, i.e. to have an officer for every relief and to leave the O/C. entirely free for supervision.

I desire to draw your attention to the fact that the German Internees saw this attempt and in fact have commented upon our general attitude in connection with British attempts to escape. I quite realise how embarrassing this whole position can be for the Government and I have endeavoured all along to ensure that the treatment for each set of Internees is identical, but while the two sets of prisoners are held so close together there are numerous little incidents occurring which give one or the other side the feeling that there is preferential treatment. While such is not really the case it can be readily understood that both sets of men are continually on the watch to ensure that one set does not get any better treatment than the other. This creates an element of continual suspicion.

Having considered the matter from all its aspects I am definitely of the opinion that it would be advisable to make the Camps distinctly separate, in fact, I feel that one or other of them should be removed outside the Camp. It might be a convenient time to consider whether or not a military camp of this nature is the most suitable place for the internment of belligerents.

Perhaps the whole question could be considered at an early date with a view to deciding the points at issue.

I have the honour to be, Sir,
Your obedient servant.

COLONEL.
(F. McNally.)

OFFICER COMMANDING: CURRAGH COMMAND.

bers of the gentry to determine who among them might be reliable. It was clear from what Covington had said that even if the internees were lucky enough to get out, the only way they'd reach the North would be through a safe house, an offer of assistance from a friendly party. While enlisting the assistance of local people was clearly a breach of their oath, so was spending their waking hours on parole planning an escape.

As good a place as any to start was Toggs Freeman, and/or Major Mitchell, mother and father of Susan and Ann respectively. They had both offered their services, but Jack knew immediately that there might be strings attached, so he was reluctant to follow up. The next obvious party was Mrs Angel Maudlins of Naas, who, in the wake of publicity surrounding Covie's April attempt, had offered her services – in a note buried in a cake. Unfortunately, the guard in charge of security for the two camps, Captain Guiney, had found it – only after the bastard had eaten the cake, or so claimed the English officers when the matter was brought up one night in the bar. The directions, which read as follows

> If you get another chance to escape, call here and
> will do best to help, One mile Naas on main Dublin Rd,
> left hand side. House covered in Trellis work. Red Roof,
> Glass conservatory, Stables, farm Buildings and Green House
> opposite side of road. All best for Easter.
>
> Burn this please.

Courtesy Irish Military Archives

were clear enough, the only problem being that, given their clarity, the Irish would have little difficulty finding them. And she was rumored to be a little dotty, one of those Empire-building expatriates, according to Shaw, who had had a few too many sherries over the years. So clearly Mrs Mullens would not be suitable.

The best choice, in Jack's view, something he had told Keefer but not the other two, at least not yet, was Major O'Sullivan, who had had the two Canadians over for dinner one night over Christmas, together with a friend of his, John Kearney, the Canadian high commissioner. Major O'Sullivan had a large estate in Foxrock, overlooking Dublin Bay. While Jack didn't want to involve Mr Kearney, necessarily, he knew that Major O'Sullivan, a retired British Army Officer, was well connected. In other words, if they did get out, it

'B'Camp 'K'Lines - British Internees.

Sir,
I have the honour to forward herewith copy of communication found in a Cake sent to the British internees by Mrs.Angel, Maudlins, Naas, Co.Kildare.
G.2 Branch, Curragh, are dealing with the matter.

I have the honour to be,
Sir,
Your obedient Servant,
Captain.
J.Guiney.
Commandant:No.I Internment Camp,
Curragh.

Courtesy Irish Military Archives

would be unlikely that the RAF would send them back. And Jack wouldn't have minded meeting the Major's daughters, either.

Finally Jack interviewed Colonel McNally, as he was obliged to. The interview went well enough, focusing on Irish history and Jack's book, the line that Jack had fed the colonel to explain the typewriter. The former reporter let it drop that he had learned a great deal about the place since they last spoke, noting his trips to Dublin and the General Post Office, scene of the great Republican uprising of 1916. He also mentioned Naas, which meant "assembly place" in its original Irish Gaelic form, Nas na Riogh, formerly the centre of one of the four Irish kingdoms, Ui Dunlainge, prior to the Norman conquest. Naas was marked by a motte – a mound that had been the site of a castle – in the middle of town. He and Keefer had stood there for a while, he said, kicking the ground. The colonel seemed impressed. And then there were the four corners of the Pale, which Jack knew all about now, being part Irish and all. The post-Parnell Renaissance world of William Yeats and George Bernard Shaw, the West Brits, the Horse Protestants, as they were known, who, like the anglophones of Quebec, had come to colonize a foreign land, and for whom time was finally running out, like characters in a Molly Keane novel. All of which the colonel readily agreed with. Indeed, the colonel had encouraged Jack to continue his studies, promising to review his work, hoping that one day the young reporter might get to see the rest of the republic, or the Irish Free State, or the Dominion, or whatever it was back then.[20]

Naturally, Jack kept trying to steer the conversation to such matters of mutual concern as staffing, overtime, camp relations with Parkgate and the adjutant general, the suitability of having two camps side by side in such an unusual setting, whether they'd soon be released, and the all important question of whether the guards would now shoot them if they tried to escape again, given what had just happened with him and Covington. But he didn't get much out of Colonel McNally there. The colonel was too smart for that.

As for Keefer, his job in the next few weeks was to coordinate the escape, to assemble materials required for a ladder large enough to span both sets of wire, and, with Fleming's help, to organize the sergeants. The first decision they made was to invite the NCO's into the officers' bar, starting at 20:00 that night, to make them feel part of the team.

"Over my dead body," said Pilot Officer John Shaw. The English officers, clearly, were not amused.

"This time Fleming's gone too far," said Welply, "NCOs in this sacred hollow? No sir, not on my mother's grave."

Even Covington, whose desire to escape was never questioned, was taken back. "I knew all along that Fleming was a communist," he said, only half joking.

It was 19:30 hours, and the ever class-conscious Englishmen had thirty minutes to stew about it.

"Welcome to the officers' bar," declared Covington with pursed lips half an hour later, attempting to make the best of it. As the first stream of NCOs arrived, Covington's stomach – with its daily diet of

20 Strictly speaking, the South of Ireland was part of the United Kingdom until 1921. Then it became the Irish Free State until 1937, Eire until 1949, and the Republic of Ireland to the present, though by its constitution the South has always claimed sovereignty over the six counties of the North. Sometimes even the Irish can't agree on this one.

NCO's in the sacred hollow, February 1941. Fleming, the thoughtful host, appears second from the right. Courtesy Bruce Girdlestone

Galloise, Madeira, and Irish stew – churned. Before long, however, he had busied himself about the bar, playing the thoughtful host.

Only a third of the NCOs showed up that first night. For one thing, they were just as uncomfortable invading the officer's turf as the officers were in having them there. The only thing that brought them through the door was the liquor, their only access to alcohol at camp being whatever they were able to smuggle past Fitzgerald at the parole hut. And for another, they weren't sure about Fleming, either.

It did lead to some useful brainstorming later on, however, particularly among the Canadians Keefer had chosen to assist him. They included Sergeant Roswell Tees from Thorold, Ontario, called Strip, who had force-landed a Hawker Hurricane in county Meath; Freddy Tisdall, a navigator from Moncton, New Brunswick, who had crashed his Hampden into the Glenadown mountains, county Donegal; Chuck Brady, a gunner from Toronto whose father was Irish, who had landed with Paul Webster, a non-commissioned pilot from Vancouver,[21] in the Irish Sea near Schull, county Cork; and, finally, Duncan Fowler, another non-commissioned pilot from Victoria, British Columbia, who had force-landed his Spitfire near Clogher, county Donegal. Fowler had been chosen for his Diston cross-cut saw (such being the benevolent nature of Ireland's neutrality, that Commandant Guiney was prepared to let a saw in but not a cake), which Keefer thought might be useful for cutting up

21 It always struck Keefer's audience as odd that there were non-commissioned pilots in charge of entire crews and washed-out pilot officers serving as navigators or gunners. It never struck the RAF that way, however.

From left, Flt Sgts Chuck Brady, Paul Webster, and Doug Woodman. Woodman, who died in hospital after the crash, was buried in Mallow with full military honours in a ceremony attended by Sir John Maffey and John Kearney. Brady and Webster were refused permission to attend. Courtesy C. Brady

pieces for the ladder; Webster, who had a degree in chemistry, for his pyrotechnical skills should they require some explosives as a diversion; Brady and Tisdall to hold up the ladder, which was bound to be unwieldy; and, Strip Tees, who had been an amateur boxer before the war, to protect them from the guards should things get a little out of hand.

As for the design and construction of the ladder, Keefer had always thought himself a bit of an expert with ladders, having worked his way through university as a lineman for the Bell Telephone company in Montreal,[22] hoisting ladders and slinging cable. He'd been looking for an opportunity to show Fleming his stuff, too, to show the flight lieutenant that he was no shirker, either.

In meetings both inside and outside camp, the four of them discussed the details over and over again. The ladder, it was decided, would have to be an unusual one. The first wire barricade was approximately ten feet high and wide, meaning that the ladder would have to be built in two sections, one for the ride up and another for the ride across, acting as a bridge. They could then jump down – since that's how most of them had arrived in the first place. This meant that one section would have to be longer than the other, to allow for some angle on the way up, and would be joined to the other section

22 Before graduating from McGill in 1939 and disconnecting permanently from the phone company.

by shackles, or nuts and bolts, assuming Keefer could find them. As for a diversion, it was agreed that an explosion in the southeast of the compound should confuse any reinforcements arriving from the duty hut. A potpourri of possibilities were discussed in Webster's presence, and he impressed them to no end, discussing detonation velocity and brisance large enough to blow up the entire camp, with blasting caps, mercury, tetryl, and trinitrotoluene.

"Where are we going to get that kind of stuff?" asked Jack.

"Easy, sir," replied Webster. "Ever heard of starch, gum, sugar, shellac, saltpetre, sir? We used to make Roman candles with that stuff all the time. Catherine wheels? Pastilles? Skyrockets?"

"Yeah, yeah, okay, Webster. Don't get carried away. We don't want to blow up the whole county for chrissake. We just want a diversion. Okay?"

"Okay, sir, no problem. I'll take care of it."

And they knew where they could find plenty of saltpetre – boil down the Irish stew.

As for the physical component, it was agreed that they should tie up as many policeman or soldiers as they could, moving fast, as they'd only have two minutes at the most. Two internees returning from parole would bind Captain Fitzpatrick's hands to the supporting beam in the middle of the police hut; hopefully, the captain wouldn't die in the explosions that followed. By then, there should be twenty or thirty internees in the outer compound. The first three would cut a trail with wire cutters through the lower wire, and the remaining ones would deal with the guards.

"You can grapple with them," Fleming ordered. "Just don't hurt them."

As the deadline got closer, Keefer focused hard on the ladder. Fortunately the Irish Army had just finished building two structures on the base, one a sports complex half a mile down the road, and the other an extra hut in their compound for increased accommodation, given all the new arrivals that fall. He had been able to collect and store most of the required materials: iron bedposts, door hinges, shower rods from the main bathroom (long enough to serve as vertical pieces), scrap timber, duckboards, and some abandoned electrical wire. The wire, doubled over, would serve for the rungs.

What they lacked was a strong wire cutter, and some bolts to somehow hinge the ladders together.

On the day they chose for the escape, 9 February, 1942, Keefer awoke early. The agreement was that everyone would take parole as usual, to avoid raising suspicion that something was up.

He had solved the last remaining problem the previous day on a trip to Newbridge. There he and Jack had found a wire cutter and a dozen used marine shackles, which weren't exactly what he had in mind but would have to do. They had brought them back into camp in their golf bags – the cutter slipped down beside Keefer's five iron, and the shackles where their golf balls were normally kept. Fortunately Captain Fitzpatrick wasn't much of a golfer, so he failed to notice. Keefer had then spent the entire day in the bicycle shed assembling the ladder, while Fleming, Girdlestone, and Calder kept watch.

All the preparations had therefore been completed as night finally fell. The moon and stars were nowhere to be seen, and the observation tower was blanketed by thick cloud. In short, it was a perfect night for an escape.

The temperature was crisp but not uncomfortable as thirty of thirty-three RAF internees assembled at 18:45 hours in the NCOs mess. Most had dressed warmly in anticipation of a successful dash to the north, with toques and gloves to protect them from the wire. A few had even cut strips from mattresses and tied them to their shins and forearms for greater protection. They had painted their faces black with shoe polish, except Remy, who had chosen to wear his fencing gear and mask instead. The three absentees were Baranowski, who had failed to return from parole, and Virtue and Wolfe, who were also on parole – only they were part of the plan.

"All right lads," said Fleming, "everyone remember their jobs?"

Despite the nods of confidence, their commanding officer took them through it once more.

"When Virtue and Wolfe come through the main gate, that's our signal. Webster and Diaper, you two head for the fence and get ready. As soon as they reach the parole hut, you launch the fireworks. Bobby, you lead the way with the wire cutters. Make sure you cut the wire wide enough to get us all through the first barrier by the parole hut. Girdle, Jack, and Covie, you'll be carrying the ladders, and Maurice, you help Bobby bolt them together, and put the bloody sword away, would you please, Maurice?"

"Oui, mon capitaine."

"The ladders get bolted together only after we've passed the first wire barrier. Make sure you let Webster up as soon as he arrives; he's got the ether and that may be useful." Webster, who looked particularly ready, had been given authority to stupefy whoever crossed his path. "The order up the ladder should be Keefer first with cutters to slice the wire, then Webster with the ether, followed by me, Remy, Girdlestone, Calder, Covington, and the rest of you. Wolfe and Virtue will be the last. "We're only going to have one minute to clear the compound so you others be quick. Diaper, you have your smoke canisters?"

"Yes sir!" Diaper replied, patting his pockets with his gloves.

"Now remember, those of you carrying the ladders fore and aft will run single file through the hole Keefer cuts. Then you'll set up side by side in the area midway between the corner gun posts – and Bobby, these ladders better hold."

"No sweat. We'll be outta here in no time," said the former Bell Telephone lineman, confidently.

A number of the internees were stuffing food rations into small back-packs, while others were adjusting their gloves and hats, anxious to get started. Most had made plans to be hidden by girlfriends or local families in stables until the army had lost interest trying to find them – however long that took. Everyone understood that once he was beyond the main gate, he would be left to his own devices.

"And, finally, don't take any extra pokes at the guards," Fleming announced. "I'm sure they'll be looking for an excuse to use their new clubs." No one had failed to notice that in the last week the guards had been sporting new clubs.[23] "If we rough them up too badly, the RAF will just send us back." He checked his new watch. "It's 19:00 hours. Virtue and Wolfe should be returning from parole in about five minutes. Let's head out." Everyone prepared to leave, talking excitedly in groups.

"Not everyone at once," Fleming interrupted. "Remember, it's just another evening. Webster and Diaper, you leave now and head over towards the eastern wall. Keefer and I will wait outside for Wolfe and Virtue. The rest of you stay here until Calder gives you the signal."

23 As requested by Colonel McNally in his 8 January 1942 letter, see pages 132–3. The internees would soon discover whether the guards had been given "definite authority to use them vigorously."

"Now listen up." Jack described the signal he would give to start the escape. "As soon as Wolfe gets within four or five steps of the parole hut, I'll yell 'lock up the Yankee swine.' When you hear that, gather at the cycle shed. From there, it's a quick dash to the first barrier – as soon as you hear the fireworks." The cycle shed was a small building beyond the guards' line of vision. Jack repeated, "Remember: when you hear 'lock up the Yankee swine,' everyone runs to the cycle shed."

Fleming and Keefer left the others and headed out across the compound. With the wire cutters stuffed into his pants, Keefer walked gingerly towards the northwest corner of the compound by the police hut. A few minutes later Wolfe and Virtue appeared, pushing their bicycles in from the main gate. They were smiling and joking. As they approached the parole hut, Jack yelled out the signal that greeted Wolfe's return: "Lock up the Yankee swine!"

"Glad to oblige you," yelled Captain Fitzpatrick in reply, opening the gate to welcome Wolfe.

Shadows could be seen running for the cycle shed. As they entered the parole hut, Virtue quickly dropped some change on the floor. As he crouched down to pick it up, he lunged at Fitzpatrick, tackled him, and quickly taped a gag to his mouth – apologizing as he did it. Flashbombs and smoke canisters began exploding at the opposite corner of the camp, surprising the internees as much as the guards. They made an impressive diversion, followed by screeching whistles and spotlights and the sound of barking dogs.

At exactly 19:15 hours Keefer set out for the wire. With Fleming one step behind him, he quickly crossed the compound, knelt down, and clipped the strands from the bottom up. The first half of the ladder, carried by Remy and Girdlestone, was rapidly bearing down on them.

"Watch out!" Keefer yelled – but it was too late. Remy, his vision obscured by the mask, had tripped and fallen while Girdlestone, the carrier in the rear, had pushed on relentlessly, driving the section of ladder into both Keefer and Remy and then into the barbed wire behind them. The two internees were knocked to the ground and pinned underneath.

After he had scrambled to his feet, Keefer looked up and saw the guards leaving the duty room. They had reacted quickly, with reinforcements already arriving from the base next door. Had they been tipped off? Eight or nine were dashing across the compound with clubs drawn. The guards ran past them in the darkness, rounding the northeastern corner and then heading south towards the smoke that was still rising in the southeast corner of the compound. Keefer finished cutting and Fleming peeled the wire back to create the opening – the diversion had worked! Remy and Girdlestone then carried the first section of the ladder through, laid it on the ground, and waited waiting for the others. More shadows moved out from behind the cycle shed. A few seconds later Calder and Covington appeared carrying the second section. Soon, both sections were laid out side by side and hinged together, each team bolting the rusted shackles into place for the ascent.

The first three internees – Keefer, Webster, and Fleming – began climbing. Keefer was the first to reach the top, swinging the second horizontal section into place. He turned to view their progress. Most of the internees had made it through the first barrier and were lined up, ready to start up the ladder. It was then that the major weakness of the plan occurred to him. If one of the shackles joining the pieces together were to break, the entire structure would collapse: everyone would be left sitting in no man's land, waiting for the guards to show up. But he knew that the ladder wouldn't break. It was the handiwork of an experienced Bell Telephone lineman!

As he set out across the bridge, he saw Webster, Fleming, Remy, and Girdlestone behind him. The ladder appeared to be holding. He scrambled to the end of the bridge and jumped, landing in a perfect parachute roll. His knee held!

A few seconds later, Webster, Remy, and Girdlestone had also jumped. Guards had begun rounding the northeast corner of the compound, racing in their direction and yelling. He looked up to see Fleming sitting on the bridge, preparing to jump, with Calder, Covington, Fowler, and Strip Tees, starting up the first vertical section. He counted heads. There would soon be an even dozen internees in the outer compound, and only eight or nine guards. They had a chance.

And then, incredibly, the ladder collapsed. The vertical piece now bearing most of the weight – fashioned with painstaking care from left-over shower pipes, duckboards, and electrical wire – had broken. The rusted marine shackles retrieved by their resourceful and supportive Newbridge merchant from the waste bins of the Dublin Yacht Club had failed them.

Without the support of the vertical piece the bridge immediately sagged into the barbed wire. Calder and Covington toppled to the ground, piling on to their colleagues beneath them.

There was no time to think. Only five of them had a chance to escape and Keefer was one. He thought of running for it but as the guards approached Webster pulled out his ether and ran towards them, brandishing the bottle as if it were a knife or broken beer glass. How Webster expected to stupefy nine guards was anyone's guess but he stopped, suddenly, screaming at the top of his lungs, "take a whiff of this you black bastards!"

"He's Canadian, he's Canadian," yelled Fleming, jumping down from the wire and landing nearby. By now, the guards were pummelling Webster with their clubs and feet, yelling anti-English epithets, like "aye, you brave defender of Libya, you're in for it now," and "I don't give a fuck for international law – do as you're told," and other such things.

Keefer ran to Webster's defence and began fighting with the guards, taking blows from their batons, shielding Webster's head and face. Fleming joined him a second later. The guards finally let up and Webster staggered to his feet, bleeding, too stunned to anesthetize anyone.

The guards then turned their attention on Keefer, but the former football star was too fast for them. He took off towards the main gate, straight-arming two corporals along the way, feinting and dodging his way through another half dozen tacklers in a dash to freedom that would have made his old coach proud. In the process he drew fire from the observation post. He kept running. He swore later that bullets grazed his head, and that it was only his speed and quickness that saved him.[24]

Just before he reached the main gate, a guard exited the duty room with a fire shovel and took a swing at his head – clipping his right ear and sending him head over heels to the ground. He was then restrained and led back towards the barbed wire. When he arrived he saw Fleming, now coiled in a fetal position, on the receiving end of Lieutenant Kelly's baton. Keefer ran to assist Fleming, swinging wildly at Kelly with both fists. He landed several blows to the head, and felt damn good about it. Eventually Kelly backed off,

24 Though this is not what the official reports would suggest. Once more they were blanks.

and Fleming was allowed to get up. No sooner had Fleming dusted himself off, however, than Webster had taken off, drawing more fire from the observation posts and additional soldiers from the main gate. They soon collared Webster, who was dragged back into the compound.

Fleming, Webster, and Keefer were taken to the police hut and locked inside. Webster and Keefer were both bleeding, and Fleming was semi-conscious from Kelly's club. Girdlestone and Remy were nowhere to be seen.

Meanwhile, the other internees were letting the guards have it.

"Kelly, why in hell were you clubbing Fleming," yelled Jack, from the other side of the wire. "We've got three or four guards in here, and more than a few pieces of iron to settle the score, so back off you stupid bastard."

"You know that none of us like the English." Kelly replied calmly.

"But most of us aren't English."

"Then don't act like the English," Kelly yelled back.

"What the fuck does that mean?"

"Don't fight in their war," Kelly replied.

Jack wasn't about to let that comment go unanswered. "You Nazi piece of filth. If it wasn't for us, you'd be singing "Deutschland Uber Alles" just like them. You'd probably like that, you yellow bastard."

"I don't think so, Calder," Kelly replied. "I'm not crazy about the Germans, either.

"If you don't like 'em, why do you talk to them so much?"

"Because I'm the only guard who speaks German. And why have you started this silly little fight with them?"

"A silly little fight? This is a fucking world war! Do you know what that is Kelly? A world war? No, I guess not. You wouldn't know a world war from a goddam potato rebellion."

Two more guards entered the fracas. Jack took a moment to compose himself. Maybe he could reason with them. "You know, Canadians aren't that different from you," he announced to the three of them. "My ancestors fought the English. And I'm proud of fighting beside the English now, I can tell you."

"Then your ambitions can't be very high," retorted one of the guards, Corporal Reilly.

Covington joined in the fracas. "Kelly, the English have been feeding you dirty cross-breds for the last 500 years. If it wasn't for us, you'd die of hunger."

"Aye, and if it wasn't for your planters, Mr Covington, we'd be a rich country. What we got instead were the Black and Tans."

"Oh, I see. Well, come over here and I'll give you the same as your Black and Tans."

"Sorry, Mr Covington. We're a free state now."

"Free all right – free to join the fucking Nazis."

And so on.

A few minutes later Girdlestone and Remy were captured. Remy, who was no longer wearing his mask, had been beaten about the head and needed hospitalization, as did three or four others. They were all accounted for. No one had escaped.

Looking out towards the western fence, their German neighbours were cheering like crazy. Those with limited English were urging the guards to use their clubs. Others were throwing rocks. To them, this was a far more satisfying show than the Calder/ Covington incident. All of the RAF internees were involved, many were bleeding, and none had escaped. They wouldn't have missed this for the world.

A few minutes later Colonel McNally arrived on the scene and asked to speak with Fleming. They spoke for a minute and then Fleming addressed the internees in an attempt to restore order. Everyone was milling about the compound, cursing and swearing revenge on the guards. Fleming didn't have much success.

Eventually the guards moved into the RAF camp and things got worse. Kelly was isolated in one corner of the yard and surrounded on all sides. Fleming ordered everyone to return to their huts. No one moved. Baranowski, who had arrived late, made amends by dragging Kelly into the middle of the yard, pinning his arms by his side, as Calder and Covington threatened to kill the Irishman. Even Fowler, normally gentle and mild-mannered, who took his frustration out on trees rather than people, was asking Fleming for permission to punch Kelly. Remy was standing nearby, accusing Kelly of clubbing him.

At that moment, the other guards arrived, wielding clubs. But they were too late. Baranowski punched Kelly first, then Remy, and finally Fowler, a devastating right-cross to the jaw. Fleming was quick to point out later that he hadn't given any of them permission.

The guards rushed to Kelly's defence. Fights broke out among several groups. Fleming yelled to Colonel McNally that the only way to diffuse the situation was to get Kelly out of the yard. Colonel McNally promptly ordered him out. Just as he was about to leave, however, another fight broke out involving Strip Tees and a couple of guards in the corridor between the police hut and the mess hall. A number of internees rushed to Tees' aid.

In the ensuing standoff, Covington yelled in the commandant's direction: "Get your fucking rabble out of our yard, McNally!" There was a suspense-filled silence. Some of the guards had moved towards Covington and others were standing by McNally, all brandishing their clubs. And then armed soldiers appeared in the yard with Ensigns at the ready, opening and closing their chambers, leaving the chilling impression that they were no longer content to shoot blanks. Indeed, it seemed as if all of the frustration on both sides had compressed itself into one moment – a word, a gesture, a flinch, or maybe just a look – whatever it took to mark the outbreak of the next Anglo-Irish war.

But nothing happened.

Colonel McNally walked across the yard, and calmly instructed the guards to remain where they were. He then turned back towards the others. "Mr Fleming," he said in a calm voice, "please order your men back to their huts. Lieutenant Kelly," he added less politely, "take the guards out to the duty hut – on the double."

The moment had passed.

Thirty minutes later five guards returned to escort Webster, Remy, Girdlestone, and Keefer to the military hospital at the other end of the base. As they crossed the outer compound, the guards asked the four internees to sign parole for the trip. They refused. Keefer hoped that an opportunity to escape might arise and refused for that reason, but in truth he couldn't have cared less if he had been on parole at that point. He was too angry. The others appeared to have reached the same conclusion. So the guards had no choice but to take their chances.

The four internees were loaded into the prisoners van and driven a couple of miles down the road, past the gymnasium to the military hospital. As they were exiting the van, sure enough, each internee took off in separate directions, cursing and yelling anti-Irish epithets over his shoulder.

Keefer was tackled almost immediately. Girdlestone dashed back up the road towards the gymnasium, but was apprehended a minute later when a number of Irish army soldiers came to the assistance of the guards. Remy didn't fare any better – he was caught and clubbed on the top of the head. By the time they reached the hospital, the four injured internees were resigned to failure. They were treated that night for their cuts and then returned to camp.

The following morning, when the adrenaline level had subsided, an uneasy tension took its place. Colonel McNally announced at a meeting in the yard that an inquiry would be ordered, and that the Irish, British, Canadian, French, Polish, and German diplomatic representatives would meet in Dublin – with de Valera – to discuss issues arising from the events of the previous day.

A short while later, when Covie headed out to the main gate for his morning cast in the Liffey, the internees were shocked to learn that their parole had been suspended, indefinitely. They were real prisoners now.

9th February, 1942.

TO:

Adjutant General,
Department of Defence
Parkgate, DUBLIN.

SUBJECT: Attempted Escape - British Internees.

Sir,

At about 19.20 hours today the British Internees made an attempt to escape. This could be described as an attempt in force. The plan evidently was as follows. Two internees engaged the policeman on the gate by a ruse and one of them made a dash to climb over the wire at the gates. The policeman on duty intercepted him while another policeman raised the alarm. At the same time a large number of the remainder of the Internees dashed for the eastern defence wires. There were two policemen on duty on that side of the compound and these were overpowered by the Internees, one of them being tied to a light standard with a flex wire and the other held down by three Internees. The remainder made an effort to get across the wires and about five of them succeeded. In doing so they made use of a policeman's sentry box which they pushed against the wire. They also made use of improvised iron ladders which were thrown across the top of the defence wires and over which they clambered to the other side. In passing I desire to state that these ladders appeared to be made from the angle sections of iron bed-steads. I am sending one up as a sample. From their appearance I imagine that they have been made outside and were brought into the Camp in sections. Luckily enough the alarm had been sounded quickly and a number of policemen had got round to the actual scene of the escape practically at the same time as the Internees were crossing the defence wires. There were several scuffles and from what I can gather there were quite a number of blows struck on both sides. Some of the British Internees were injured but none seriously. Some of the policemen also complained of injuries. All the Internees were returned safely to the Compound. Subsequently the injured Internees were treated at the Hospital. I attach a list of the injured. As may be observed most of the injuries were due to barbed wire and none of them are serious.

I arrived at the Camp between 19.35 - 19.40 hours and the situation at this time was fairly serious. The British Internees were truculent and abusive. I went into the Officers quarters accompanied by Lieut. Kelly (Officer on duty in the Internment Camp). I found the British Internee Officers very resentful and they complained very bitterly that they were badly treated, in fact they accused our men of holding down their Officers while they beat them. During this time Captain Baranauski who had been on parole returned. He struck Lieut. Kelly a blow on the nose causing him to bleed. His cowardly

For the enquiry that followed the 9 February incident, over fifty reports were submitted to the adjutant general in Parkgate. No testimony was allowed. Reproduced here in their entirety are four reports, those of Colonel McNally, Commandant Guiney, Grant Fleming, and Jack Calder. Bobby Keefer refused to write a report, thinking it was a waste of time. He turned out to be right. Courtesy Irish Military Archives action was apparently in conformity with the feelings of the remainder that were there. I might state here that they appeared particularly bitter against Lieut. Kelly. This Officer had of necessity to use force in order to prevent P.O. Girdlestone getting over the protective wires. I am satisfied that if he had not taken this action Girdlestone would have escaped. The Internees were very slow to settle down but at about 20.30 hours order had been completely restored.

I desire to draw attention again to my previous communication on this matter and to suggest that it is time that some definite ruling or guidance would be given to the matter of dealing with situations of this kind. The situation is practically impossible. It must be borne in mind that there are 34 Internees and it takes at least 3 or possibly 4 men to hold each internee and lead him to the quarters without injuring him. I have not this number of men to spare and it can be readily understood, therefore, that the police must use some considerable force when a general effort of this type is made.

I presume that there will be repercussions as a result of the injuries to the British Internees and consequently I beg to suggest that a Court of Enquiry should be convened and I feel that in all the circumstances it would be better if this Court was convened by you. The question of whether or not the Internees should be permitted to give evidence at this Court should also be taken into consideration.

I have the honour to be, Sir,
Your obedient servant.

COLONEL.
(T. McNally.)

OFFICER COMMANDING: CURRAGH COMMAND.

Copy to Chief of Staff.

REF.IC/94/40.

NO.I INTERNMENT CAMP,
CURRAGH CAMP,
10th.February,1942.

CONFIDENTIAL.

To/
Provost Marshal,
Dept.of Defence,
Parkgate, Dublin.

Attempted escape of British Internees from 'B' Camp on 9th.February,1942.

Sir,

I have the honour to inform you that at about 19.15 hours on 9th instant, three British Internees approached the Cage apparently with the object of going on parole. One of them - Sgt.Virtue - was admitted to the Cage and as soon as he entered he dropped money on the ground and stooped to look for it. Just then the sound of the policeman's whistle was heard from the Compound and at the same time Virtue,who was in the Cage attempted to climb up the outer gate but was caught by the policeman, another of the internees,who was waiting to be admitted to the Cage,started to climp up the wire,while a third tried to tie the gate evidently with the intention of preventing help from entering the Compound. During this time a number of internees rushed out of one of the huts towards the Eastern side of the Camp, some of them overpowered the two police and the N.C.O.,on duty in the Compound. One man was tied with wire to a Post while the other two struggled on the ground. During this time some of the internees cut through the single wire fence which surrounds the buildings, they then cut portion of the main defences and with the aid of ladders the following succeeded in getting out over the wire:-

Flight Lieut.Fleming,
Pilot Officer Keefer,
Sergeant Webster,
Flight Lieut.Remy,
Sub.Lieut.Girdlestone,

They were subsequently captured by the police.

Two iron sectional ladders, one seventeen feet the other eleven feet, were used by the internees. These ladders were apparently made outside and brought in and assembled by means of bolts and nuts. Each section was [illegible] inches long when bolted together,and the width of the ladders was one foot.Another ladder was made from lengths of wood taken from the newly constructed hut, the rungs of this ladder were made from lengths of electric flex. Two duck-boards from the bath-room were bolted together and also brought into use to get over the wire. They also used smoke generators, one of which did not go off and which we hold presently.

In the melee a bottle was broken in Sergeant Webster's pocket, which gave off fumes like that of ether,apparently intended for the policemen when overpowered. Also "Squibs" were thrown at the Police which some complained had the effect of dazzling them.

While five of the internees had succeeded in getting out over the wire,a number of them were between the wires and on the ladders when the police got around.

A fracas developed, the P.As had to use force to restore order, i.e.,use their batons. Together with this those who had got outside resisted returning to Camp resulting in force having to be used also. The force used was not excessive in my opinion.

The following were treated in the General Military Hospital for minor injuries:-

Flight Lieut.Remy,
P/O. Keefer,
Sub.Lieut.Girdlestone,
Sergeant Webster.

Camp 1, referred to in the second page of Guiney's report, was where the IRA prisoners were kept. As noted by the Irish officer, the possibility of IRA-AF complicity, considered by some of the RAF internees at Jammet's restaurant in Dublin as early as the fall of 1941, was also on the minds of their gaolers.

They were taken to the Hospital under escort as they refused to sign the Parole when requested to do so by me. When emerging from the cars at the Hospital Sub.Lieut.Girdlestone bolted, got away down the road towards the Gymnasium and in front of this building he was recaptured. He resisted violently and had to be restrained by the Police

The injuries caused were chiefly cuts from barbed wire. The internees were very arguementative and were using language which was very provocative and disrespectful in character towards the Irish Race generally,i.e.,"there is a smell off ye - ye dirty Irish", "ye yellow bellied curs,ye're like the Italians" and characteristically Anglo-Irish in bitterness etc.

The Command O.C.,was early on the scene and is conversant with the whole affair. Myself and Comdt.O'Neill were subsequently speaking to F/O.Ward,at the latters request. He said among many things they would use Clubs on the next occasion. He also stated that if Lieutenant Kelly came into the Compound they would not be responsible for their actions,using very forcible language against him. Lieut.Kelly was struck on the face by Flight Lieut.Remy and Captain Baranowski while accompanying the Command O.C.,Colonel McNally, into the British Officers' Quarters.

I would like to mention in this report that the situation in "K"Lines is becoming extremely difficult. There are certain aspects in its relation to this Camp that are causing me great anxiety, for instance last night and on other occasions all available Officers and Police off duty, including the "Stand To Party" had to be rushed to 'K'Lines to deal with the situation that had arisen there. It is all very well while nothing happens in No.I on occasions like these,but supposing there did,we could very easily be confronted with a very grave situation. It is not outside the bounds of possibility that a joint arrangement could develop on those lines, i.e.,supposing the 'C'Internees started a racket and the I.R.A., were conversant with the fact,then this problem would present itself in all its seriousness.

I am,therefore,of the opinion that 'B' and 'C' Camps should be divorced from No.I,with an independent adminstration, and another arguement in favour of the seperation is that the character of the adminstration of both Camps are entirely different.

Herewith,copies of reports from Lieut.J.A.Kelly and the Police concerned.

I have the honour to be,
Sir,
Your obedient Servant,

J.Guiney. Comdt.
Commandant;No.I Internment Camp,Curragh.

11th February, 1942

Flight Lieutenant Fleming's Report.

I have the honour to report that at 19.15 hours on the 9th of February, 1942, an attempt to escape was made by members of the British Internment Camp, excluding Captain Baranowski who had not at that time returned from parole.

None of the Internees carried a weapon and all were under orders by me not to use violence on the guards. A smoke canister was to be thrown outside the main gate and I instructed Sgt.Diaper to throw it some distance away so that it would not hit or fall near the guard or guards at the gate. Sgt. Diaper informed me that he had carried out these orders. Before going over the centre double barbed wire entanglement I saw a guard free so held his arms to his sides until I was relieved by others.

Flying Officer Remy (France), Pilot Officer Keefer (Canada), Sub-Lieutenant Girdlestone (New Zealand), Sergeant Webster (Canada), and I (Canada) managed to get outside the double barbed wire. I was caught for a time on the top outside of the double entanglement and saw the rough treatment Sergeant Webster was receiving and objected in strong language but made no reference to nationality of guards. I managed to drop into the lower wire apron where Sergeant Divers and three or four others held my arms and beat and clubbed me to the ground where I was kicked and clubbed until almost senseless. Apparently some of the guards were intended to beat and not hold us as they carried two batons - one each hand. During this time I was being called a British Bastard, dirty yellow English and many other filthy names; also there were many remarks such as "Where are all you defenders of Libya"? "Where is England now"? This shows the unneutral and anti-British attitude of a large number of the guards.

While I was getting a further beating Lieutenant Kelly rushed up swinging a baton menacingly but when I called to him that he had better hit me while I was being held, he left the group. Sgt. Webster who had been caught and dragged back to the centre fence was again beaten for protesting at treatment of me. There seemed to be no discipline among the guards but some of the Internees informed me later that the Sergeant who had just received his promotion was trying to restore order among the guards. Someone swung a shovel at Pilot Officer Keefer but missed him. He was treated reasonably well when the guards heard he was a Canadian. During the whole period approximately fifteen or twenty shots were fired by guards.

When all the internees had been forced into the mess hall and officers' mess, Colonel McNally, O.C., Curragh Command, and Lt. Kelly came into camp preceded by approximately twenty guards carrying batons. I accompanied Colonel McNally to the door of the officers' mess where I was roughly pushed and pulled aside by guards until a Sergeant Major, Eire Army, pushed roughly into the officers' mess after the Colonel and Lieutenant. I gained the mess with difficulty and saw Captain Baranowski threatening Lieutenant Kelly, so I held Baranowski's arms and quietened him. Colonel McNally expressed disbelief of the behaviour of the guards but told us he would hold an enquiry into it as he did not want this sort of thing.

At this time some of the guards in the hall struck two British sergeants apparently with a view to starting a fight but were unsuccessful as the Sergeants refused to retaliate. Colonel McNally immediately sent his guards out of the camp. I attach the reports of Sergeants Tees and Fowler, both Canadians.

I requested a Medical Officer who came immediately to the camp and Flying Officer Remy, Pilot Officer Keefer, Sub-Lieutenant Girdlestone and I returned under escort with him to the hospital. On alighting from the van Sub-Lieutenant Girdlestone escaped from the right front door and ran out of sight down the road pursued by two guards and an officer. I next saw him being brought back to hospital by the two guards who were beating him with their free fists. The officer was making no attempt to stop this treatment. I had been detained in the van and asked to see Girdlestone but was cursed by the guards at the front van door. I then requested to see an officer but was again cursed and told to shut up by guards at the back door. A tall Corporal who had been one to help bring back Girdlestone rejoined the others at the rear of the van saying "The fucking bastard" I wish I'd broken his God-damned neck'. The others made it quite plain that they agreed with him.

Flying Officer Remy had been thrust into the van and when he objected to treatment of Sub-Lieutenant Girdlestone the guards at the rear tried to provoke him to come out for some of the same.

After sitting in the van for 40 minutes I was treated in hospital and returned to camp.

A short while after my return to camp I saw Sergeant Webster collapse and caught by two or three others who helped him to his room. Fearing concussion I asked again for the Medical Officer although he had already examined Webster. When he arrived Webster insisted that he was all right and that only his arm was bruised so the Medical Officer examined his arm and left camp.

The reader will note in both the reports of Fleming (this page) and Calder (next page) some of the offensive language. Other examples can be found in other reports not reproduced – but similar in spirit to some of Covington's salutations earlier.

11th February, 1942.

Statement of Pilot Officer J. Calder, R.C.A.F., re Attempted Escape from B. Internment Camp.

For the attempt to escape on Monday, February 9th, 1942, I was in charge of the groups detailed to handle guards who approached the ladders erected over the main barrier. These groups were instructed to carry nothing in the way of weapons and to try to impede the guards by grappling only. It was decided to tie these guards to posts where possible.

These instructions, so far as I was able to observe, were carried out implicitly. At one point Sgt. Fowler asked for, and was refused, permission to use his fists on a guard because some of our men were being clubbed outside the second fence.

At the zero hour I notified the men in the main huts and in the first and second huts to move. By the time I got over the first fence, two guards were being held near the ladders. The ladder I was to have used collapsed beneath the weight of Flight Lieutenant Fleming and Sub-Lieutenant Girdlestone, and I shouted to them to use the other ladders. When Flight Lieutenant Fleming tackled and held a third guard who approached, I relieved him and held the guard while Fleming climbed over one of the ladders. I told Pilot Officer Keefer and Pilot Officer Midgeley to follow him. The guard I was holding managed to raise his baton, which I grabbed and threw out of the way into the compound.

Several Sergeants came to my aid and I was then able to see what was going on outside the main barrier. Scores of military police were there. I saw Flight Lieutenant Fleming struggle down the barbed wire into the crowd of corporals, where he was held and beaten with clubs. I can identify one corporal who was beating him from behind.

I saw Pilot Officer Keefer jump down into the group and struggle through, although the guards attempted to club him.

Flight Lieutenant Fleming was finally knocked to the ground where the clubbing was continued. I heard shouts of "Dirty English" from the guards. Sergeant Webster was subjected to a severe mauling on the ground.

I attempted without success to get up one of the broken ladders. Then I shouted through the fence that unless the beatings stopped we would use pieces of ladders to slug the guards inside the fence. I used words to the effect that the Irish could not be proud of the fact that they were beating up men who had helped to defend Eire, and that our men had sought only to go back and fight in England.

About this time I saw Lieutenant Kelly of the Eire Army away from a group outside the fence in which Sub-Lieutenant Girdlestone was being held. Lieutenant Kelly's hair was messed and his face was flushed.

"Did you see the clubbing and slugging that your men were doing out there, Kelly?" I demanded.

"Yes, and I'm proud of them", he called back. "You'll all be treated like that".

The milling around outside broke up soon but several of us refused to leave the area between the first and second fences for several minutes. A corporal whom I can identify tried to push me back into the main compound.

"I'm not going for a while yet", I declared, "I'm pretty damned sore."

"You're just going to start more violence if you stay here", he said. "You know, these men don't like the English".

"But I'm not English", I said, "I'm Canadian, and some of my ancestors fought the English in this country. But I'm proud of fighting beside the English, if that's what you're getting at".

"Then your ambitions can't be very high", he retorted.

I went to the mess and soon afterward Colonel McNally arrived, accompanied by Lieutenant Kelly and a large body of men, still armed with batons. Several officers were discussing the incidents with Colonel McNally when I heard a commotion in one of the corridors. I went out and heard that Sergeant Tees had been struck without warning by an Irish corporal.

I told Colonel McNally to get the corporals out of the mess before they started further trouble. I believe I used the words "rabble" and "filth" in this connection. Colonel McNally ordered them out.

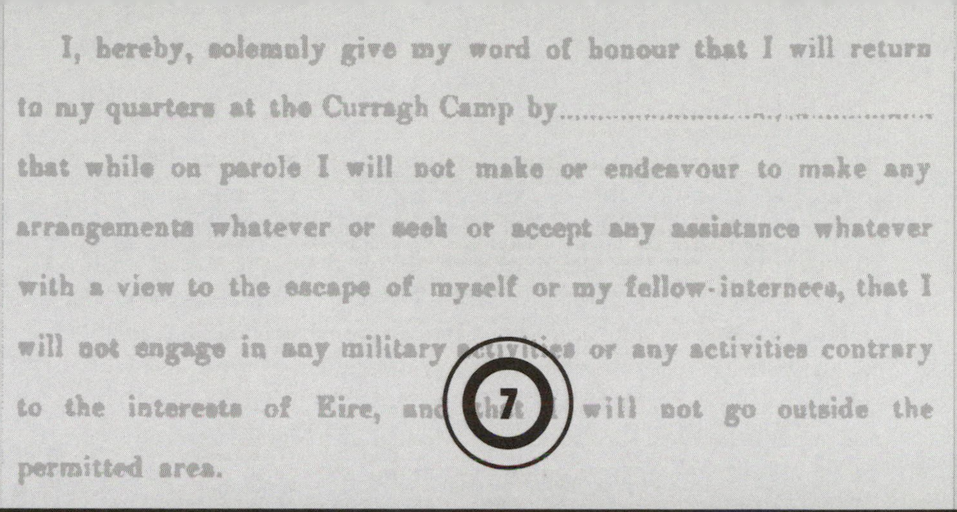

I, hereby, solemnly give my word of honour that I will return to my quarters at the Curragh Camp by.................................. that while on parole I will not make or endeavour to make any arrangements whatever or seek or accept any assistance whatever with a view to the escape of myself or my fellow-internees, that I will not engage in any military activities or any activities contrary to the interests of Eire, and that I will not go outside the permitted area.

7

WAITING FOR TIPPERARY

The aftermath of the February 9th incident provided plenty of opportunity for reflection. For the guards, for the internees, and of course for Jack Calder, a flyer trying to make the best of a rotten situation and yet a reporter whose job it was to investigate, assess, and report the truth as he saw fit – if that's what he saw fit to do. And with parole now suspended, and so much time on his hands, why not? Why not set the groundwork for resuming a brilliant young career, telling all who cared to hear of their unfortunate situation. Surely it beat hanging around the compound all day with the others, chewing on a twig.

If one were to record and understand recent events, Jack decided, the reaction of the guards to the recently concluded escape attempt was as good a place to start as any. Particularly its violent nature, which seemed to engender a certain delight on their part, as if the vanquished had suddenly risen up and defeated the oppressor after so many years of slavery and subjugation. He had, after all, told Lieutenant Kelly and the others whom he had steadfastly refused to smack that his ancestors had fought *against* the English and that should have made *all* the difference. And yet it hadn't. The guard had clubbed him anyway. Why did he do that? Was that the fighting Irish spirit Jack had heard so much about?

Nothing like a little spirited debate to focus attention on the matters at hand.

While the answers to these questions were obviously not simple ones, Jack assumed that one explanation for the antagonism lay in religious differences. If it was obvious to him that the close relationship between the internees and the local people – the wealthy Horse Protestants of the Pale – was a major irritant, it was nonetheless a surprise. As the son of an Anglican minister, Jack was attuned to the issue of religion, of course, but not when religion was so tied up with history, and money. In Protestant Ontario it was never an issue with the Irish-Canadians he'd met, but in Ireland in the 1940s it was. And, when he thought about it, who wouldn't be bitter? While he hadn't seen much of the place beyond the Pale, they all knew that while the Kildare Hunt Club was busy with its weekly parties at Ardenaude, or Osberstown House, or Ballynure Eustache in Wicklow where his friend Ann Mitchell lived, the rest of the island was rationing tea, half an ounce per week, per person. As they, the internees, were eating beef every day, others were living on soup and potatoes, trains were stopping in the middle of the night, children were out scrounging for coal.

And it hardly helped that the internees on full RAF pay were making double what the guards were.

And then there was the matter of the South's religious affinity with the Germans – though Lieutenant Kelly and others would deny it, no doubt, if Jack was ever able to interview them on the topic. Still, the shared Catholicism must be at least *partly* responsible for some of the "un-neutral" comments – Libya and Italy aside. He knew that when the Germans had attacked Sudentland, de Valera had told the English that Hitler's actions to "protect" the Catholic minority in that part of the world were no worse than if he, de Valera, had sent an army into Belfast to protect the Irish Catholic minority in the North. The English were contemplating conscription in the North at the time, while offering to end Partition if de Valera chose to declare his support for the Allies. That seemed curious, of course, not so much that Ireland would refuse Chamberlain and Churchill's offers of a united Ireland in exchange for joining the Allied cause[25] – that wasn't curious, that was treachery – but because the Irish leader would actually compare himself to Hitler,

25 While these negotiations are described in a number of post-war texts, one comprehensive account can be found in Tim Pat Coogan's biography of de Valera (*De Valera, Long Fellow, Long Shadow*. [London: Arrow Books, 1995, 548–54]).

even embrace some of his policies, a tactic that hardly enamoured him with the rest of the free world. So clearly there was some Irish sympathy towards Germany, Jack realized, even if the sympathy was kept in check by the latter's excesses.

And then there was name calling too, which was silly, yet very real. Jack had seen that in Europe, of course, where violent words had already led to violent deeds. He'd also seen it on the sports beat of the Chatham *Daily News* covering high school hockey, the analogy being as if some idiot on one team had stood up on the bench and yelled, "we've been feedin' you dirty cross bred Irish pigs for five hundred years" (that would have been Covington) with a second idiot on the other team skating over and yelling back "ye stupid English bastards; where are ye brave defenders of Libya now?" (that would have been Corporal Smeaton,) and then all of a sudden both benches had emptied and everyone had been tossed. Libya? Boy they really knew how to hurt a guy. Rommel had just taken Tobruk that month, and the British 8th Army's counter-offensive at el-Alamein was still weeks away.

In addition to history and religion as root causes of this antagonism there was politics, of course, the root of *all* evil, especially at that point in history. Jack would definitely have to acquaint himself with Irish politics, he realized, if he were to understand where the guards were coming from – or write anything more about the place than he had to date. And that would be a job and a half since understanding Irish politics, like understanding Irish religion or history, can drive anyone crazy. He knew, as they all knew, that de Valera was not a flash in the pan; that the Irish leader had been around for a generation already, that as a member of the IRA de Valera had taken over as leader of Ireland immediately after the Americans had saved him from execution in an English jail during the last war; that he had then resigned during the treaty negotiations which followed after a majority of the Irish parliament had sworn allegiance to the Crown. What seemed important to some Irish – that former colonies, by their oaths, retain some semblance of loyalty to the regal benefactors – seemed less important to de Valera. There had been a civil war after that, Jack knew that much; he had heard all about the Black and Tans well before the insults started flying back and forth with the guards.[26] Still, how and why a disagreement over an oath can lead to people getting killed demanded scrutiny – even if oaths can do that to people sometimes, make them do crazy things – whether oaths of allegiance to an unwanted sovereign, or parole oaths.[27] He knew that the guards all supported neutrality and scoffed at the supposed support for the alternative championed by the opposition party, the fine gael, so maybe that was another factor. After de Valera had been defeated on the issue of the oath, and had been forced to resign his membership in the IRA, he had formed his own party, the fianna fail, which had gone head to head with the fine gael, or the United Ireland Party (UIP), a party formed by a fellow named Cosgrave, with men like James Dillon, whose name appeared in the *Irish Times* from time to time, playing a supporting role. Covington and a few others had met Dillon, Jack learned, and were impressed. The fine gael was backed by wealthy farmers, businessmen, professional people, and landowners, including the Horse Protestants of Kildare, which certainly couldn't have helped. Jack realized if he were to get to the bottom of all of this, he would have to interview as many politicians as he could on his next monthly trip to Dublin.[28]

Not that he could really do anything about it, or at least publish anything he managed to produce, without a great deal of effort. Censorship was rife during the war, especially in Ireland, as de Valera was never a big fan of the press anyway, describing the *Christian Science Monitor* in May 1918 as the "devil's chief agency in the modern world." Still, Jack wanted to understand what Mr Kearney had told him and Keefer over dinner on their last trip to

26 Covington's reference to the black and tans at page 146 – "come over here and I'll give you the same as your black and tans" – was taken verbatim from the archives. Black and tans were auxiliary police officers from England who were employed by unionist forces against the republicans after many Irish police and army officers refused to fight their countrymen. Infamous for their savagery, these officers owed their name to the colour of their uniforms – light khaki coats and dark trousers and caps. A black and tan is also a variety of beer and/or bloodhound, depending upon the context.

27 Not that the former reporter would ever write about Wolfe – or ever did. As Keefer put it, when asked, that would have been awfully close for comfort.

28 Jack Calder's attempts to engage James Dillon, among others, in this debate are referred to in the archives and inferentially in his second article on Irish neutrality, which appeared in the *Toronto Star* and other Canadian papers on 10 June 1943. Once more, for the impatient reader, reference is had to page 226.

Dublin, something that the Canadian High Commissioner had repeated on his next visit to camp the following week: that a neutral county among belligerents needed censorship, presumably to protect their tenuous position. Nonetheless, did they really have to ban *The Great Dictator* and the *Spy in Black*,[29] Jack wondered, without fully understanding the zealous, almost puritanical, nature of de Valera, the man the English commentators and editorialists had derisively dubbed "the Playboy of the Free State," after he had attempted to block the Abbey Theatre's production of Yeats' best known work in America some years earlier. In addition, why would the Irish government allow Heir Heinkell, the German ambassador, to fly the swastika over his residence in Dublin, while at the same time denying Sir John Maffey, the British representative, the right to fly the Union Jack over his?[30] As puritanical and zealous as de V was, the anti-British feeling was undeniable. Memories of the civil war were still fresh, and the only British troops or airmen anyone wanted in Ireland were the POWs, kept tightly under lock and key. No, there was no doubt about it. De Valera and his ilk really didn't like the English.

Which brought him full circle really, for while he wanted to understand all of this, Jack knew his opportunities would be limited. Particularly with G2, the intelligence arm of the Irish army led by Commandant MacKay, now checking his mail in and out, steaming open his envelopes and copy-typing his telegrams. One afternoon later that month, Colonel McNally called him into his office, and finally explained why.

The former reporter had been out in the courtyard, exercising with the others. The colonel was at his desk, writing reports, as usual.

"Busy, Colonel?" asked Jack, politely closing the door behind

29 The internees, however, could see whatever they liked at the camp cinema since the censorship rules, at least for films, didn't apply to them. Either Maffey or Kearney brought them in through diplomatic channels once a month.

30 Some surmise that this example of preferential treatment may not have been an accident. There had already been two German "mistakes" to that point in the war: in August 1940, when Nazi bombs killed 3 in Campile, and then again in May 1941, when 34 others perished in Dublin – quite apart from the intentional bombing of Belfast which resulted in 700 deaths. By the following year, it's fair to assume that the Germans had made their point and the Irish were anxious to avoid any more "mistakes." (The English, on the other hand, would never have attacked, hence the double standard.)

him, and taking a seat.

"Yes. And I understand you have been busy too, Calder."

"Colonel?"

The camp commandant offered Jack a cup of tea, which Jack accepted. "Writing," he replied, "that's what I'm referring to." Jack assumed the colonel was referring either to his own report on the escape, or his soon-to-be-written book on Irish history – the line that Jack had fed the colonel to explain the typewriter. He thought the report was fair and balanced, as one might expect. As for the book, the colonel, and Jack, had yet to see any of it.

"I've been contacted by the Canadian high commissioner," the colonel continued, "Mr Kearney in Dublin, who tells me that a magazine in Canada wants to publish a story you wrote last year about your landing in Claire. *"I Flew into Trouble."* It's a large Canadian magazine, I don't recall the name ..." The colonel paused to search for a document on his desk.

"Sir? I don't know what you're talking about," Jack replied indignantly. The former reporter was delighted. It had to be *Macleans*, Canada's national news magazine! Excellent! But how did he find out?

"*Maclean's*. And don't play silly bugger with me, Calder," the colonel snapped, irritated. First Covington, and now Calder; they didn't pay him enough. "Now, I don't mind if you write while you're here. That was your living before the war and I assume it will be after. Life is difficult for you, I appreciate that. You're bored, I appreciate that too. And I want you to feel as comfortable as possible. But you told me it was Irish history you were writing about, not Irish politics." Jack flinched. "Now, frankly, I don't see any harm in it, but next time I want to know. Got it?"

Jack was taken back. The colonel had figured it all out. He knew everything, but he wasn't mad. Why? What should he say?

"GOT IT!"

"Yes sir, Colonel," Jack replied a moment later.

So the former reporter certainly had a lot on his plate in the wake of the big fight, figuring out where their differences lay, why they couldn't just get along. Fortunately, he had a lot of time to ruminate, sitting around camp, staring at the wire, shuffling the dirt beneath his feet like a real prisoner. He had a lot to absorb – the religion, the history, the politics – and now this, now that his big secret was out. Why did the Colonel seem pleased? He doubted very much that the RAF, or the other internees, would take the same view. Would he get beat up too?

For Keefer the waiting was interminable. Not just during the period they were confined to camp, which, depending on the source, lasted somewhere between a few days and a month,[31] but even after that, when the Irish government decided to do nothing, ignoring Commandant Guiney and Colonel McNally's warnings, choosing to leave the camp where it was and sending them back out on parole instead. The months that followed seemed to stretch on endlessly. Waiting for the RAF to make up its mind, having first told them that it was their duty to escape and now telling them that the situation in Ireland after Pearl Harbor was too critical, that they would be better served by staying put. Waiting to see if there were any more arrivals, and finding out that they weren't; that the internees were windowdressing just as Remy had said. The honour of Eire demanding that they be locked up, and their own honour keeping them there. Waiting to see his family and friends back home, and his pals in Lincolnshire, Slapsy Maxie, Billie O'Brien and the others, who were telling him to hurry up and bust out, that he was missing all the fun. Waiting to see the girls again, the English ones, who made up for in enthusiasm what they lacked in good looks, as Slapsy used to say. Waiting for Picadilly, for Lester Square. Waiting for war, remembering it as even more fun than Ireland.

So he began writing too, not like Jack, but simple letters home describing the highlights, hoping that his mother would at least get the general idea, given some of the rumours circulating about the place. No, they weren't going to the movies with the Germans, sitting beside them in the dark, drawing their weapons when the lights came back on and the credits rolled. No, they weren't going to church together, sitting in the same pews, listening to the Reverend Flint urge greater understanding of their differences.[32] No, they

31 Records kept at the Irish Military Archives hardly clarify the issue. One report suggests thirteen days, another thirty days, and a third, three. Keefer couldn't recall, other than a general impression that it must have lasted much longer than that.

32 As if they should be expected to settle their differences when the Irish could hardly settle theirs. Still a pleasant thought.

Life here is like (imagine) being marooned on a desert island with Hedy Lamarr – okay for about the first 3 days (or should I say nights) and then damned boring

Another excerpt from one of Keefer's letters home. This was one of the family favourites.

weren't sitting side by side in the local pub, either, though there was that one incident, of course, but that had been a mistake. In fact he'd discovered that month that Oberlieutnent Neymeyr, the parachutist he had met in the basement of the Osbterstown Hotel on New Year's Eve, had escaped soon afterwards, becoming the first and only German to escape the Curragh which meant something to him years later if to no one else.

Ref:-S/297/P.

Confidential.

G.2 Branch,
Headquarters,
Curragh Command.

26th June, 1944.

Chief Staff Officer,
G.2 Branch,
Department of Defence,
Dublin.

Conrad Neymeyr – escaped German Internee.

Sir,

Your ref. No. G2/3441 refers.

I have the honour to forward herewith for your attention a letter written by Hauptmann Mollenhauer to above named now a prisoner at Internment Camp "44", Canada. You will notice that it is addressed to Kapitanleutnant (Lieutenant Commander) Konrad Neymeyr No. 56507, Kriegsmarine (N). It was not known here that Neymeyr was a Naval Officer though his walk certainly looked Naval, and I do not know how Mollenhauer learned that he had been promoted to the rank of Lt/Commander.

I have the honour to be,
Sir,
Your obedient servant,

(signed) COMDT.
(R. Mackey).
Officer i/c. G.2 Branch, Curragh Command.

Other records (at both the Irish Archives and the National Archives in Ottawa) establish that Neymeyr successfully jumped a cattle boat to Spain and continued spying for the Third Reich until he was eventually arrested by the British in Portugal and shipped to a POW camp in Alberta, of all places. Courtesy Irish Military Archives

The Allies were going to win the war and he wanted to be in on it.

Whenever he was feeling particularly depressed, Keefer would grab his clubs and head out to the camp links, routinely passing Captain Fitzpatrick on the way, who would force him to empty his bag at the police hut. There had been rumours floating around as to how those pesky internees had managed to smuggle the wire cutter and bolts into camp, which had apparently triggered the captain's sudden interest in the sport. He and Jack became friendly with the Lawlor twins that summer, golfing at least a couple of times a week,

The guilty golfer. Note the size of the bag, large enough to store a wire cutter and half a dozen rusted marine shackles.

Portmarnock Golf Club. Courtesy Bill Gibson (and his excellent book, *Early Irish Golf* [Ireland: Oakleaf Publishers, 1992])

rain or shine, feeling guilty all the while, jousting back and forth with their new friends, calling them bloody taigs but in a friendly way. They played beyond the Curragh too, including a round at Portmarnock north of Dublin, one of the world's great courses. There they saw Major O'Sullivan, the retired British Army officer, who repeated his offer of assistance over dinner at Foxrock. They also played bridge with Grace, the dowager, a friend of the O'Sullivans, and enjoyed the company of one of the Major's daughters, Sheila, kid sister to Maureen. Keefer remembered her photograph on the major's bookshelf, scantily clad and standing alongside Johnny Weismuller in "Tarzan, King of the Jungle." They also enjoyed a round with Major Mitchell, Ann's father, at Bray to the south, in Wicklow.

When he wasn't busy golfing, or stretching the boundaries of his parole, Keefer did his best to kill time at the racetrack, with men like Aubrey Brabizon, the trainer at the Aga Khan's stud farm in Kildare not far from the camp, who knew all the jockeys, all the hot tips.

Babby O'Keefer and Jackie FitzCalder, waiting impatiently to be released, but in the meantime doing their best to fill the time.

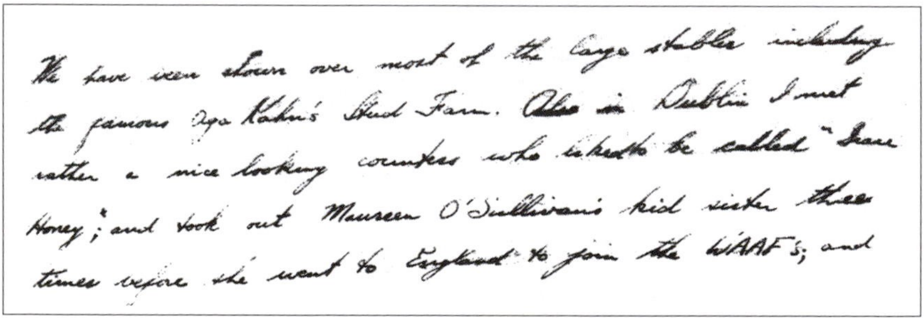
We have been shown over most of the Cayo stables including
the famous Aga Kahn's Stud Farm. Also in Dublin I met
rather a nice looking countess who liked to be called "Lace
Honey"; and took out Maureen O'Sullivan's kid sister three
times before she went to England to join the WAAF's; and

The Aga Khan stables referred to in another of Keefer's letters, above, later became the National Stud, one of the world's largest breeder of thoroughbreds. The identity of the nice-looking countess unfortunately remains a mystery.

"The IRA? What's this about," asked Fleming, staring at his empty glass of claret, Albert, the head waiter, nowhere to be seen. "It's hard to believe that they would help us."

Four of them had joined Remy at Jammet's on their monthly Dublin visit – Keefer, Calder, Girdlestone, and Fleming – so desperate that even their commanding officer was now prepared to turn to England's worst enemy for assistance.

"Not so, mon capitaine," replied Remy, explaining to Fleming, as he had to Keefer and Calder a couple of trips earlier, that, while the IRA may have declared war on England (most recently in 1938), cooperation with its allies was quite another matter. Look at America, with all the hot-bloodied republicans there sending their hard-earned yankee dollars across the Atlantic. That was exactly how de Valera, their arch enemy, had gotten started. "The only people on the island with the intelligence network and arms to get the job done are the IRA. What do we have to lose?"

As fellow enemies of the Irish Free State, maybe the RAF internees and the IRA did have something in common.

"The English officers will love us for this," Keefer remarked to Fleming as Albert finally returned with another bottle of claret. "I think it's a crazy idea."

"Maurice might be right," Jack said. "Some of the guys de V interned managed to get away with a million rounds of ammo from Phoenix Park. They certainly have the fire power. All we need to do is find someone who can set it up. I'm sure they'll want to bust their boys out too, but does that matter to us?"

Keefer cringed. For one thing, it did matter. And for another, he hated it when Jack referred to the Irish leader as de V. Like the guy was his best friend, or something.

"C'est ça," said Remy. "Before your arrival there were rumours about a German invasion when Churchill was negotiating for the ports.[33] Back at Dunkirk. They were afraid that the IRA was in on it. But they would never expect the IRA to help us."

"I see," added Fleming. "The IRA would free us first, as a diversion, and then get their boys out. And the Army would be a lot more interested in catching them than us. It might work ..."

33 In fact, the rumours referred to by Remy of German plans to invade Ireland were confirmed after the war. There were two plans: *Operation Green*, in 1940, which called for a marine assault of 4,000 troops from France along the south-east corridor from Wexford to Dungarven, and second, Hitler's personal invasion plan, in April 1941, timed to coincide with the 25th anniversary of the Easter Rebellion, a diversion to a full scale assault on Britain. England's draft response, known as *Plan W*, involved two divisions of infantryman arriving in convoys across the Irish Sea, waiving the tricolour to distinguish them from, say, Cromwell's troops in 1641. (See Fisk, *In Time of War: Ireland, Ulster, and the Price of Neutrality* [London: Andre Deutsch, 1983.])

"I still think it's crazy," said Keefer. "The IRA blazing into camp, shooting everybody in sight, risking a bloody civil war – just for us."

"I can see it. Remember, we're prisoners too. And do you want to stick around for the rest of the war listening to the diplomats, Bobby?" Fleming shot back. "There's no way that Maffey is going to get us out of the Curragh." Fleming had been far from impressed with the British representative – both how he had handled the Bud Wolfe affair, and what he had thought of the February attempt. "Why should we care about what happens to Ireland?"

Just then an attractive, redheaded waitress emerged from the kitchen.

"That, mon capitaine, is our contact."

Fleming sat back and admired the contact.

Remy met her eye and motioned her over to the table. She had a pale silky complexion and a thin, narrow face. Maurice introduced her to the crowd.

"Bonsoir, mademoiselle, celui-çi sont mes amis, Bobby, Grant, Jack, et Bruce."

The woman looked at the four of them, and then down at Maurice. "My name is May and I don't speak French ..."

"Eh, bien ..."

"And I might be a mademoiselle, but don't get to thinking I'd be interested in the likes of you. There's no way."

And they had thought the head waiter was snotty. She turned and left.

"Looks like you got the wrong May."

"No, no, Girdle. This is just part of her act."

May returned from the kitchen, slapping the rest of the meal down on the table. She scowled at Remy, but raised an eyebrow in Fleming's direction. The Canadian smiled, then leaned back in his chair. That was enough for him.

"One last thing, lads. Not a word about this to anyone. Not the Brits or the others – and certainly not Ward. And in the meantime, wish me luck with the woman named May. I'll need it."

The following week the camp saw the first successful escape since Keefer's arrival the previous fall. Once more it involved Bud Wolfe, and once more not as the American would have preferred.

The four of them – Keefer, Calder, Fleming, and Girdlestone – had been out on the camp links in the rain, and had returned just in time for the excitement. It was 16:00 hours when the American showed up at the bar, grinning from ear to ear.

"Check this out." Wolfe was waiving a cancelled parole slip in the air, in Webster's name, for all to see.

It had finally worked.

Under the new procedures put in place after the failed escape in February, parole forms, which were filed alphabetically in a drawer in the police hut, were to be returned once cancelled to the individual internee upon his re-entry into camp. Colonel McNally's orders. The Irish wished to avoid another situation like Wolfe's in the week following Pearl Harbor. Of course, most of them would leave before Captain Fitzpatrick could actually hand them back the cancelled form,[34] and the Irish hadn't tried any other means of exacting their promise, such as by slipping it under the internee's door. The internees had therefore decided as a group that those with last names next to each other in the alphabet, and thus filed next to each other in the drawer, should stagger their return to camp as much as possible just in case Fitzpatrick cancelled the wrong form. And, fortunately for Webster, his name immediately preceded Wolfe's.

"See, check this out, guys, it's signed by Webster not me. His parole has been cancelled!" The American began reading paragraph 2 below: "'The period of parole will not be regarded as terminated until the signed form has been duly returned to the signatory.' So? Get it? Get it?"

"Ya, we get it. Where's Webster?"

"He's in Naas."

"Okay, that's good enough for me," said Fleming. "Let's get the slip to Webster, fast."

They called in Chuck Brady, Webster's pilot, who fortunately was still in camp. "Brady, find Webster fast, and give him this." Fleming gave him the the cancelled form, in case the RAF wanted to see it. "And here," Fleming then pulled out five Irish pounds. (Keefer had recently noticed that Fleming always had money on him.) "That should be enough to get him across the border."

"Yessir!"

34 As noted by Colonel McNally in his memorandum following the meeting with senior government officials on 23 February. See page 111.

Brady did find Webster in a pub in Naas – sadly, as it turned out. For while Webster made it to the North alright, and rejoined his unit the following day, he died three months later off the Dutch Coast.

The form below dated 15 April 1942, is the one duly signed by Webster, cancelled, and then returned in error to Wolfe, triggering Webster's escape. The fact that it could still be found at the archives fifty years later, alongside Wolfe's "un-cancelled" form above shows what good sports the Irish turned out to be in the end, otherwise they would have destroyed the evidence. Courtesy Irish Military Archives

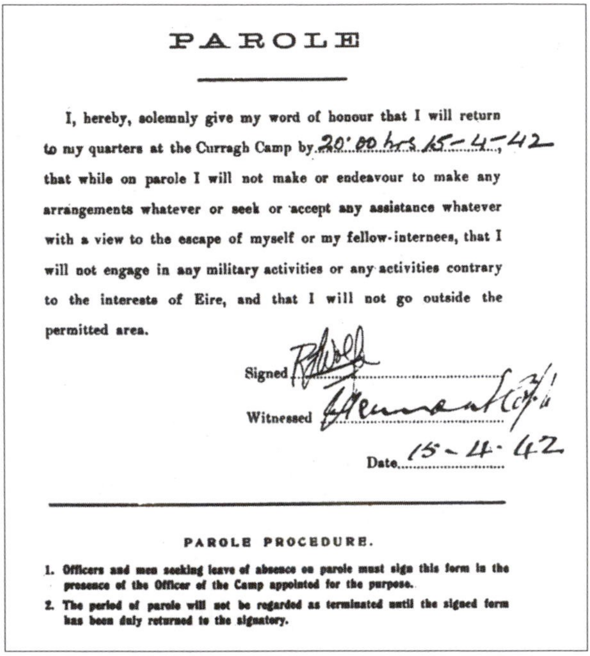

PAROLE

I, hereby, solemnly give my word of honour that I will return to my quarters at the Curragh Camp by 20:00 hrs 15-4-42, that while on parole I will not make or endeavour to make any arrangements whatever or seek or accept any assistance whatever with a view to the escape of myself or my fellow-internees, that I will not engage in any military activities or any activities contrary to the interests of Eire, and that I will not go outside the permitted area.

Signed ....

Witnessed ....

Date 15-4-42

PAROLE PROCEDURE.

1. Officers and men seeking leave of absence on parole must sign this form in the presence of the Officer of the Camp appointed for the purpose.
2. The period of parole will not be regarded as terminated until the signed form has been duly returned to the signatory.

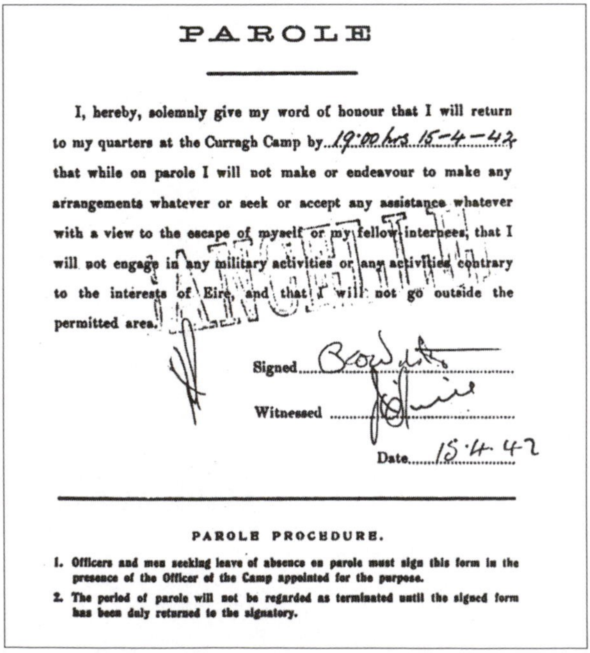

PAROLE

I, hereby, solemnly give my word of honour that I will return to my quarters at the Curragh Camp by 19:00 hrs 15-4-42 that while on parole I will not make or endeavour to make any arrangements whatever or seek or accept any assistance whatever with a view to the escape of myself or my fellow-internees, that I will not engage in any military activities or any activities contrary to the interests of Eire, and that I will not go outside the permitted area.

Signed ....

Witnessed ....

Date 15·4·42

PAROLE PROCEDURE.

1. Officers and men seeking leave of absence on parole must sign this form in the presence of the Officer of the Camp appointed for the purpose.
2. The period of parole will not be regarded as terminated until the signed form has been duly returned to the signatory.

16.4.42

O.C.
No. 1 Int. Camp

I wish to inform you that Sgt. Webster, British internee, left camp on Parole at 14.45 hours 15th instant having signed until 18.00 hours. At 14.55 hours Webster returned and said he wanted his parole extended to 02.00 hours. A new Parole Form was prepared and timed until 02.00 hours 16th and signed by Webster who then left; the earlier parole form being cancelled and retained is enclosed.

From enquiries I have made it would appear that Webster's last parole was cancelled by mistake at 15.55 hours and handed to P/O Wolfe instead of Wolfe's own parole which terminated at 18.00 hours.

P/O Wolfe went on parole at 16.15 hours having signed until 18.00 hours. He returned at 16.55 hours, this is the time we are assuming he was handed Webster's parole.

Wolfe again availed of parole at 17.00 hours to 20.00 hours and returned at 18.45 hours. At this time his earlier Parole which terminated at 18.00 hours was cancelled and endorsed 45 minutes absent and handed to him, leaving the one which ended at 20.00 hours and which should have been cancelled at this time, still on the file. It is enclosed herewith.

The Paroles are arranged in alphabetical order so that all the "W's" would be on the same file. The error was not discovered until about midnight when it was seen that there were two Parole Forms for Wolfe, one ending at 20.00 hours 15th and one ending at 02.00 hours, while there was none for Webster who was still out.

Wolfe's Parole ending at 02.00 was the third and last Parole he signed when he left at 20.40 hours, returning at 01.25 hours, 16th.

Signed, J. O'Neill, Comdt

As the careful reader will note, the *Sun* (below right) didn't quite get it right. No one actually got out of the camp during the 9 February attempt. Courtesy Irish Military Archives, and *Vancouver Sun*, 24 July 1942

Sergt. Pilot Webster Is Lost In Air Operations Overseas

A Vancouver flyer who has been shot down and interned in Eire, who twice escaped internment, is now reported missing in air operations July 24.

He is Sergeant Pilot Paul Osborne Webster, son of William O. Webster, manager here for the Underwood Elliott-Fisher Company, and Mrs. Webster of 2230 Southwest Marine drive.

SGT. PILOT. WEBSTER.

Webster's plane crashed at sea off the coast of Eire in October, 1941, killing the observer. He got his crew into a rubber boat and headed for the coast where they were picked up by an Irish fisherman. Ashore they buried the observer's body and were arrested and interned.

FIRST TRY FAILS.

Three months ago the crew attempted mass escape but was caught and returned to the camp.

A month later Webster escaped and made his way to England where he joined his R. A. F. squadron.

For the second time since their son joined the air force here two years ago, Mr. and Mrs. Webster have been advised that he is missing.

Paul Osborne Webster left the University, where he was preparing for a career in radio research engineering to join the R. C. A. F. Previously he attended Magee, and was well known in high school football.

For the next while, despite his unswerving determination to return to the war, Keefer slipped back to where he had been prior to the February attempt. Sheila O'Sullivan having moved to England, and seemingly disinterested, he began seeing Susan Freeman once more, as the rhododendrons and azaleas turned a fiery red and a man's heart turned to thoughts of spring. They would often meet at Jim and Tom Lawlor's hotel, Osberstown House, for tennis, or a cycle through the countryside, or a picnic, or golf, or whatever struck them that day, sometimes alone, sometimes in a group, but usually with female companionship. Bruce Girdlestone, the New Zealander, had taken a fancy to Cora Waddington, the daughter of Nesbitt Waddington, the Aga Khan's general manager, while Fleming had become friendly with Mary Kelly, a friend of Susan and Ann. Often a large group of them would get together.

Some relationships that spring and summer became more serious. In June, Covington announced his engagement to Nora McMahon, a Catholic from Newbridge, which caused quite a fuss, particularly when he suggested to Colonel McNally that he could avail himself of the same arrangement Ward had with his wife, moving in with Nora's parents until they married, and then finding a place of their own. The colonel promised he would consider it, but in truth Covington, with his six previous escape attempts, might not have been the best candidate. And then Keefer's second pilot, Les Diaper, announced that he was dating a Catholic girl from Brownstown, Jessica Black, while his nose gunner, Maurice Brown, who was Catholic, was dating a Protestant girl, Brigit Devlin, also from Brownstown. This seemed to confirm Jack's theory: that what separated people in the South wasn't so much religion as it was economics, though maybe it was something else, what Jack had described for *Maclean's* magazine as "lulling."

Jack's friendship with Ann Mitchell – a striking woman, who like Sheila O'Sullivan had recently signed up for the WAAFs or Women's Auxiliary Air Force – strengthened with each passing week. She was leaving for England at the end of the summer but would be back. Whether all this mating was on account of spring, envy, or the fact that Maffey had forbidden them to escape is anyone's guess.

In the last week of May, Jack's brother returned for a visit. Since his last visit Jack had been racking his brain to figure out how the

On the steps of Osberstown, from top left: unidentified woman, Welply, Mary Kelly, Fleming, Cora Waddington, Girdlestone, Sgt Tisdal lying down with the beer. Courtesy Bruce Girdlestone

Ann Mitchell. Courtesy Bruce Girdlestone

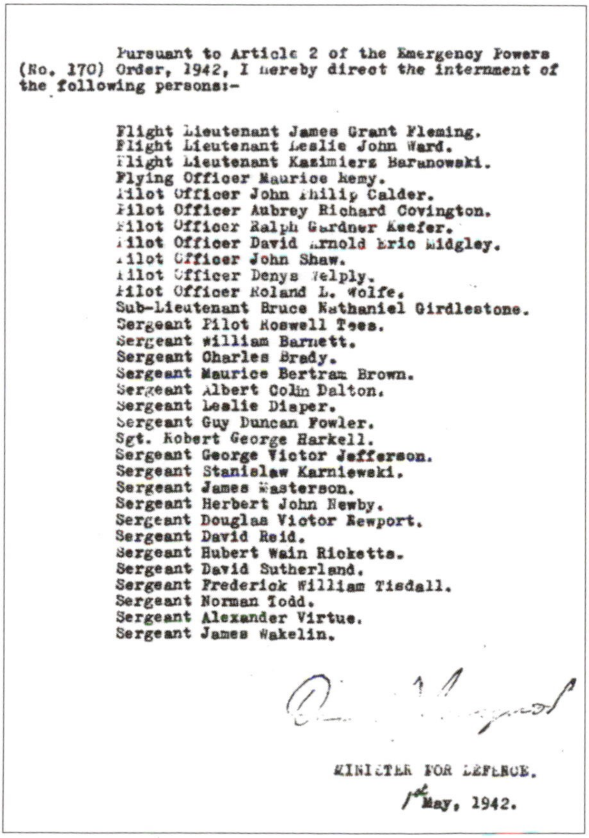

Pursuant to Article 2 of the Emergency Powers (No. 170) Order, 1942, I hereby direct the internment of the following persons:-

Flight Lieutenant James Grant Fleming.
Flight Lieutenant Leslie John Ward.
Flight Lieutenant Kazimierz Baranowski.
Flying Officer Maurice Remy.
Pilot Officer John Philip Calder.
Pilot Officer Aubrey Richard Covington.
Pilot Officer Ralph Gardner Keefer.
Pilot Officer David Arnold Eric Midgley.
Pilot Officer John Shaw.
Pilot Officer Denys Welply.
Pilot Officer Roland L. Wolfe.
Sub-Lieutenant Bruce Nathaniel Girdlestone.
Sergeant Pilot Roswell Tees.
Sergeant William Barnett.
Sergeant Charles Brady.
Sergeant Maurice Bertram Brown.
Sergeant Albert Colin Dalton.
Sergeant Leslie Diaper.
Sergeant Guy Duncan Fowler.
Sgt. Robert George Harkell.
Sergeant George Victor Jefferson.
Sergeant Stanislaw Karniewski.
Sergeant James Masterson.
Sergeant Herbert John Newby.
Sergeant Douglas Victor Newport.
Sergeant David Reid.
Sergeant Hubert Wain Ricketts.
Sergeant David Sutherland.
Sergeant Frederick William Tisdall.
Sergeant Norman Todd.
Sergeant Alexander Virtue.
Sergeant James Wakelin.

MINISTER FOR DEFENCE.
1st May, 1942.

The official internment list, published by the Irish Ministry of Defence on 1 May 1942, a requirement under Article 77 of the *Geneva Prisoners of War Convention of 1929*. Before the publication of this list, their internment was actually unlawful according to international law, an argument that someone would have come up with eventually, no doubt. Courtesy Irish Military Archives colonel had found out about the article. Over dinner at the Mitchell's in Wicklow he finally got some answers. His brother explained that Canadian Press had felt the article was too long for a newspaper but had sent it along to *Maclean's* as a favour to Jack. *Maclean's* thought that in order to include the CP file photos of the Wellingtons from their Brest daylight raid, they'd need both RAF and RCAF clearance. And, while they thought that the aerial shot of the camp was great, they also thought that they'd need Irish Army approval to use it. Finally, they thought they'd better get Canada's Department of External Affairs involved, just in case, knowing how sensitive the Irish were about such things. Mr Kearney had then stepped in and talked to someone in the Irish government, either Mr Walshe, Eire's secretary of state, or de Valera himself, and then the matter had been passed down to Colonel McNally.

As to why the colonel had seemed receptive, the impression was that the Irish would be happy to have something written about their internment if it wasn't unduly critical. They wanted to dispel some of the rumours, not only the usual anti-neutrality ones but also the ones that had started locally, early on, after Covington and Mayhew had landed, which had then been spread across the Irish Sea by a few nasty-minded English reporters anxious to make a name for themselves. Rumours of beatings, squalour, starvation, and torture which had started to bother even the internees after a while, including their commanding officer at that time who, after he'd escaped with Mayhew, had attempted to set the record straight.

Still, it got Jack thinking: what if I wrote something they didn't like? What would the colonel do then?

"The destruction of our culture, Bobby, began in 1166, during the so-called Anglo-Norman invasion, when Henry II decided to eliminate our pagan ways." Susan Freeman had begun speaking with temerity but then with growing confidence as the minutes, and finally hours, passed.

"You see, Bobby, two of our kings, Dermot MacMurrough of Leinster and Tiernan O'Rourke of Breifne, had been feuding and MacMurrough abducted Dervogilla, O'Rourke's wife."

"No wonder Henry invaded, that's serious stuff."

"Yes, well, legend has it that she preferred it that way. Anyway, MacMurrough fled to England with Dervogilla, and then

B Internment Camp
20.6.41

[Copy typed]
Dear Captain Fitzpatrick,

Thank you for showing me the letter from the Irish Red Cross Society to the Secretary of the Department of Defence, dated June 11, 1941.
I am shocked to see that such misinformed rumours are current about us here and I would be glad if you informed anyone interested that there is no truth whatever in [such] statements...
We object strongly to good-natured people who pass on rumours about us giving the impression that we are tramps or orphans and though we very much appreciate any gifts of green vegetables, cakes, honey, books, cigarettes, etc., I would like it to be known that we are not in any way short of clothes, individuals buying their own in Dublin.
I would be glad to know of the source of any rumours of this nature, so that I may put them right.

Yours sincerely,

Hugh Verity, F/O

Hugh Verity, who escaped from the Curragh later that month (June 1941) went on to write the very popular *We Landed by Moonlight* (Air Data, 1995), a first-hand account of the Lysander resistance flights into France and Belgium later in the war. Courtesy Irish Military Archives

to France, where they enlisted the help of that scoundrel Henry."

"Who at that time was king of both England and France."

"Be quiet! How can you expect to learn anything about our culture if you don't listen! That's better ... yes ... Henry II was king of both England and France as you say, and Henry went to the pope, Adrian IV, an Englishman – which was a mistake that would never again be repeated, thank the Lord – and got permission to invade our lands. But Henry changed his mind, being more concerned with France. MacMurrough then went to Wales and there he found the Earl of Pembroke – Strongbow – another scoundrel who would do Henry's dirty work for him. By 1170 Dublin had fallen to the Normans."

"You mean we only have another eight centuries to go?"

"Be patient, please!" Susan was clearly getting impatient herself. Unlike her friends Ann Mitchill or Sheila O'Sullivan, she had begun discovering her Gaelic roots, a sudden awareness that she was no longer English, or even Anglo-Irish, but Irish, plain and simple, and she felt it was important that Keefer understand. "The English passed the Poyning's Laws in 1366, making our language, laws, and customs illegal – in our own land! England controlled the Pale then."

"The eastern area of Ireland."

"You have been listening," she said endearingly. "But not the rest of our country. Henry VIII would change all of that, of course. The plantations were his legacy."

"The one with all the wives."

"Yes, the one with all the wives." She found his little boy charm appealing, she had to admit. "Henry said that the English needed our ports to defend themselves against Spain. Does that sound familiar?"

"Yeah, but Hitler is conquering all of Europe, Miss Freeman."

Keefer called her Miss Freeman whenever she got on his nerves. Unfortunately it reminded Susan of her mother.

"And Ironsides, bloody savages they were," she snapped defiantly. "Torching our monasteries, murdering our people. To hell or Connaught, what kind of choice was that! Raping our women." Tears welled in her eyes. "And then the treaty of Limerick, when our people were banned from their own government, the courts, the army, even their own schools ... bloody English bastards."

Keefer flinched.

After a pause she asked, "Do you Canadians form an independent country."

"Well," he replied, groping for words, "the Brits leave us pretty much alone." He wasn't quite sure how to answer that one.

"Do you have a constitution like us?" she asked, referring to Eire's constitution, enacted in 1937.

"Well, no, our laws are passed at Westminster."

"And why do you Canadians allow that?"

"It hasn't been a problem so far," he replied, espousing the view of an anglophone Canadian raised in Ontario. "The Americans have a constitution and God knows we're different from them."

"Eire, do you know what that means, Bobby?"

"What, Susan?"

The setting was pleasant enough, sitting under a sprawling chestnut tree on a blanket with a wealthy young heiress, near a copse of lilacs where the two had first passed on the hunt, meandering at the back of the pack, thinking nothing but romantic, lustful thoughts. Only now the blossoms were definitely withering.

"It means 'Ireland, sovereign, to be united,' and much more."

"Not a very concise language," he replied flippantly.

"And it'll happen, Bobby, de V will do it," she replied, more softly now. She saw no good in irritating him if her goal was to educate him. In the past year she had become a big fan of Eamon de Valera, a man, she believed, who had fought long and hard for an independent Ireland and had the diplomatic skills to preserve it, despite all that was happening around him.

"But what does Eire have to lose by releasing us, really," Keefer asked. He hated it when Susan called de Valera that – almost as much as when Jack did. "The Battle of Britain is over, Susan, and Ireland is not as important as you think." He quickly apologized, but by then it was too late.

Jack's next story wasn't exactly what everyone expected. Word had leaked about the *Maclean's* piece, and everyone was anxiously looking forward to that. So reading a CP wire story about the February break instead – what Keefer would later call the Ladder from Hell – definitely caught people by surprise. Especially the censors, who began by checking the date of its publication, 27 June – Jack had written the piece two weeks before that, by most, though not all, accounts – the sub-heading, "That Let Down Feeling" – likely the brain child of some wise-ass editor trying to capitalize on the lighter side of the war; it certainly wasn't Jack's – and then the piece de resistance, the byline, "Rathdrum, county Wicklow, Eire" which was his idea, knowing it would keep the G2 branch of the Army, and Commandant's Mackay, busy for months. Rathdrum, who lives there for goodness sake?

Of course, had they done any real checking they would have noticed that Ballynure Eustache, county Wicklow, was where Ann lived, at one end of the road, and Rathdrum, county Wicklow, was where the alleged Canadian Press representative lived, at the other. A remarkable coincidence. But then they wouldn't have suspected her, since Jack wasn't allowed to leave the county without special permission.

Two can play at this game, figured the former reporter.

"Wrong, gentlemen," snapped Fleming one morning the following month, when the blossoms of spring had given way to the long, hot days of July. The weather wasn't the only thing making Grant Fleming hot under the collar.

Montreal Star, 27 July 1942.

That Let-down Feeling!

Broken Ladder Foils Escape of Canadian Internees in Eire

Bob Keefer of McGill Is Possible Victim of Plot That Failed

RATHDRUM, County Wicklow, Eire, June 27 — (C. P.) — The story of a rough-and-tumble attempted mass escape by Canadian and English airmen from Curragh internment camp now can be told. It is a story of bumps, bruises and cuts; of barbed wire, ladders and wire-cutters.

The scramble and dash through the darkness of a cold night not long ago came to grief because of one thing—"Ersatz" bolts in the ladder joints. When the ladders collapsed, the guards rushed in.

Names of those who participated in the break have never been made public. But among Canadians known to be at the camp are Flt.-Lt. Grant Fleming, Calgary; P. O. R. G. (Bob) Keefer, Montreal; F/O. Jack Calder, Goderich, Ont., and Sgt. Pilot Paul Webster, Vancouver. Sub. Lieut. Bruce Girdlestone, a New Zealander, in the Fleet Air Arm, is also interned. All had made acquaintances outside the camp while on special honor leave.

EX-McGILL MAN

(Keefer is a former football player at McGill University. He and Calder, a Canadian Press editor at Toronto prior to enlistment, were members of a bomber crew forced to bale out over neutral Eire last fall.)

Stories from the camp of the attempted break say that officers and non-commissioned men alike had a part. There was marked ingenuity in their plans. Through some means they obtained small pieces of wood to be fastened together into 16-foot ladders. Lacking steel bolts, they used anything that could be found and appeared serviceable.

At their zero hour the airmen slipped safely from their quarters into the darkness. They gathered at an agreed-upon place and crept toward the first barbed wire fence. Some went to work with wire-cutting tools while others set up the ladders. They got through and over safely.

LADDER SNAPS

The next fence proved their undoing. The swaying ladders snapped when the bolts gave way and the men were dashed into the barbed wire or were sent tumbling to the ground.

As they scrambled to their feet the guards rushed up. In the scuffle and milling about in the darkness some of the fliers broke loose and headed for another fence. But more guards came on the run and the men were caught. They were marched back to their quarters and all were accounted for.

Residents on The Curragh over the hills to County Kildare say the men's punishment was confinement to quarters for two or three weeks.

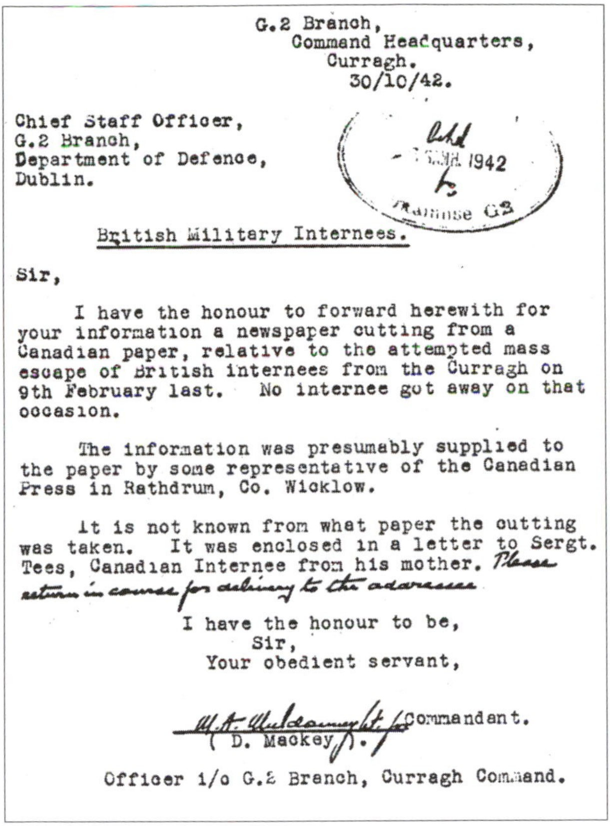

G.2 Branch,
Command Headquarters,
Curragh.
30/10/42.

Chief Staff Officer,
G.2 Branch,
Department of Defence,
Dublin.

British Military Internees.

Sir,

I have the honour to forward herewith for your information a newspaper cutting from a Canadian paper, relative to the attempted mass escape of British internees from the Curragh on 9th February last. No internee got away on that occasion.

The information was presumably supplied to the paper by some representative of the Canadian Press in Rathdrum, Co. Wicklow.

It is not known from what paper the cutting was taken. It was enclosed in a letter to Sergt. Tees, Canadian Internee from his mother. Please return in course for delivery to the addressee

I have the honour to be,
Sir,
Your obedient servant,

Commandant.
(D. Mackey).

Officer i/c G.2 Branch, Curragh Command.

Handwritten note below reads as follows: "Spoke to Captain McCall. He's of the opinion that no representative of the Canadian press is resident in Eire. Comdt. Smith's opinion is that the information was passed to some person in N. Ireland for transmission, the Rathdrum address being merely subterfuge. Capt McCall concurs with this opinion." Courtesy Irish Military Archives

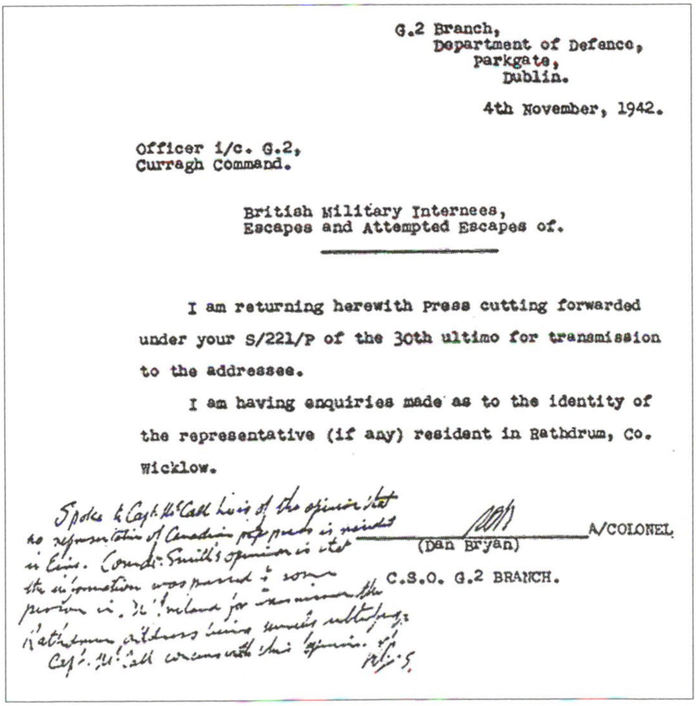

G.2 Branch,
Department of Defence,
Parkgate,
Dublin.

4th November, 1942.

Officer i/c. G.2,
Curragh Command.

British Military Internees,
Escapes and Attempted Escapes of.

I am returning herewith Press cutting forwarded under your S/221/P of the 30th ultimo for transmission to the addressee.

I am having enquiries made as to the identity of the representative (if any) resident in Rathdrum, Co. Wicklow.

A/COLONEL
(Dan Bryan)
C.S.O. G.2 BRANCH.

"Well then, sir, what did this woman say?" Shaw asked.

Somehow the British internees had found out about May, and were none too pleased.

"Yes, old boy," added Welply, "do tell us what you and this Irish trollop have been discussing these past several weeks. Was this conversation before or after dinner?"

"Now shut the hell up, all of you," their CO replied. "Her friend will have armed support, with half-a-dozen soldiers. They'll arrive at night in a large lorry. They'll drive the bloody thing through the cattle gate, smashing it to pieces."

"What's his name?" asked Shaw, a little more interested this time.

"I don't know. He's from Tipperary. They'll split an opening through the inner wire. We'll pile in the back, and if any of the guards try and stop us, his men will open fire over their heads. No one will be shot. That's the agreement."

"How do we know that?" asked Shaw.

"And how are we going to get to the border?" asked Covington, doubtfully.

"May says that her contact will have that arranged," continued Fleming. "He has an inside guy with the army who can tell us where the road blocks will be. They'll either bribe the soldiers at the block, or shoot their way through."

"An inside guy in the army?" asked Covington, suspiciously.

"Yeah," Fleming replied quickly. "Once we get through the first few roadblocks, we'll take the back roads to the border. There's no way the army will catch us."

"Unless we stop in a pub," said Jack, looking over at Covington.

"Shut up, Jack. Now, this is top secret. Lives will be at stake and I want no one talking about this plan, particularly with Ward, or anyone else who's likely to tell Maffey."

"Doesn't he know?" asked Midgely.

"No, Midge, he doesn't," replied Fleming. While Maffey didn't know, Fleming had approached Wing Commander Begg the previous week. He realised after meeting May on his last trip to Dublin that he would need some more money: at 500 pounds a head, he didn't have any choice. To his surprise Begg had approved the plan but warned Fleming that he would have to let the English officers in on it. Begg suggested referring to the contact as a friendly insider with the Irish army and leaving it at that.

"Grant, tell us more about this contact," Covington asked, a little more persistent this time. Something smelled, and although the bar officer's liver had been taking a beating, his nose still worked fine.

"Well, like I said, he's from Tipperary." Fleming was hiding something and Covington knew it.

"This man from Tipperary," he asked, "what does he do for a living?"

"I don't know ... odd jobs I guess. He's got some kind of a business, but he's struggling."

"He's not IRA, per chance?" demanded the senior English officer.

Fleming looked nervously at Remy, then at Keefer. They're going to find out sooner or later. "Indeed, he is, Covie, indeed he is."

All hell broke loose.

"You've asked those bastards to help us," Shaw stammered, increduously.

"The bloody IRA?" said Midgely, aghast. "You Canadians are more stupid than I thought."

"Those savages are dangerous," Shaw continued, "They'd sooner line us against the wall and empty their chambers than help us."

Welply was too shocked to say anything.

"You're right, Midge, that's probably their plan," Shaw suggested. "Once they've killed all of his majesty's officers they'll pocket the reward and our friend from Tipperary will spend it at Osberstown House, buying drinks for the Nazis."

"Bloody animals," said Midgely.

Welply was now watching Fleming suspiciously. "Well perhaps it's not a bad idea, chaps." He said guardedly. "Exactly how long have you been associated with them, Mr Fleming?"

Fleming looked impatiently at Welply and sighed. "All my life, Denys. You've got me lads. I don't see eye to eye with my brothers about the war, but I agree with them that all Englishmen are assholes."

"I told you, John." Welply quickly looked over at Shaw.

"For Godsake, Welply, shut up." Covington interrupted in a firm voice, commanding silence from his colleagues. He had waited through the last exchange but was now prepared to address the issue. "Tell us more about this, sir," he asked formally, stopping Shaw with a cold stare and granting Fleming the respect he was due.

"Unless we do something about it, gentlemen, we're stuck here for the rest of the war." Fleming was speaking quickly now, grateful for the support. "Shaw ... Midgely ... and you, Welply ... you can all wait for your beloved Sir John to get us out of this mess, but it won't happen. It's not in Maffey's interest to help us." He was snapping his words quickly now with an intensity that allowed no interruption. "As we all know now, we're the windowdressing for the British government's policy here – there's no avoiding it. Why else are they releasing the new arrivals, but keeping us?" Fleming stood up and crossed the room, letting his words sink in. They were still listening.

"If we escaped," continued Fleming, "it would upset the apple-cart. They won't allow it. Kearney told me last week that he, Maffey, and Begg will visit the camp next week 'to explain things' Why? Ask yourselves! You probably won't," he added sarcastically, "but I'll tell you why. They're talking about moving us to a camp north of Dublin – with no parole." Fleming emphasized his last point by sitting down at the bar and tapping his whiskey glass three times on the oak. Although he had lied to scare them – about the no parole part, the rest was true – he felt it was justified. How else could he light a fire under these men? Especially, when he, himself, was beginning to tire.

Each of the officers swallowed hard, staring back at Fleming.

"Did you mention your idea to Begg?" Covington asked.

"No I didn't," Fleming replied, lying once more. "I wanted to bring it up here first." He turned quickly to the three British officers, seated together at the bar. "I knew you'd think I was crazy – John, Midge, Denys – but it's the only way. The IRA can help us if we're careful. We might not see eye to eye with them about a lot of things, but if they can be of some use to us, who gives us a shit?"

Welply turned away in disgust, taking a long draught from his pint.

"The boss is right, boys," Wolfe declared. "Those fuckers wanna keep us here. They didn't have to send me back. I tell you, my unit sure wanted me. Those poor buggers are getting shot to hell. They need us."

Fleming noted Wolfe's new enthusiasm. He was pleased.

"What's in it for the IRA if they help us escape?" Covington asked.

"They want five hundred pounds for each internee," Fleming replied.

"Where are we going to get that kind of money?" Wolfe asked.

"We'll find it," Fleming assured them. He had been startled by the speed with which the wing commander had come up with the money.

"The RAF will court martial us for this, you know," said Shaw.

"No they won't, John. You shouldn't assume that Begg agrees with Maffey and the politicians."

Fleming had assumed as much, at least until he'd met the man alone, until he had come to understand the reality of their situation.

"You're crazy, Fleming," Welply announced, unconvinced. "If a wing commander of the Royal Air Force agreed to reward thugs from the Irish Republican Army for whatever purpose, then the wing commander of the Royal Air Force should be court-martialled. Isn't it enough that they're bombing our cities, killing innocent women and children?"

"We're making a living out of bombing German cities, if you haven't noticed, Welply."

"That's different, we're at war with Germany." There was a cool, defensive air to Welply's retort.

"Well, they're at war with you, aren't they?" The others were flabbergasted.

"Their war is not an honourable war, sir," snapped Welply. "And if you proceed with this plan, then I'd sooner resign my commission than accept assistance from those savages."

"That'll save us 500 pounds, then won't it," Fleming replied.

"How long are we going to have to wait for this guy anyway," asked Keefer one morning, as the three Canadians sat in the sun, poles in hand, along the banks of the River Liffey, watching the trout go by.

"I don't know." Fleming had seemed morose the last few days, worrying his countrymen, who had come to trust him.

"You don't know?"

"No, I don't know."

"What did she tell you last time?"

"She told me ... um ... she told me ...

"Grant?"

"She told me ... that she was waiting to hear back."

It was the last week of July, the week prior to the Irish Derby, and they were still waiting. And it wasn't doing Grant Fleming any good. He was getting in over his head, he knew it. First there was Mary Kelly, daughter of the anglo-Irish ascendancy, and now there was May and the IRA. And, of course, conversations like this one – short, aimless, exchanges which left each of them scratching his head, wondering what all the fuss was about. They could just as easily stay there. He was losing it! Grant Fleming of all people!

His first sight of her – tall, striking red-hair, slender, pale face, intense blue-green eyes – he'd known she spelt trouble. He could think only of her now, waiting for him in her seedy east side Dublin flat. He knew that he wasn't acting as a war hero should, but then he had never been in a situation like this before. He could never tell Keefer or Calder. He was responsible for them. He was their commanding officer.

"So, how are you going to get the money to him?" asked Keefer, watching the fish go by. The trout in the Liffey were plentiful back then. He'd catch the next one.

"Who?

"The man from Tipperary. Christ, Grant."

"Sorry, Bobby. She suggested that I leave some money under a plate at Jammet's as a show of good faith."

"Which plate," Keefer asked.

"What?"

"The plate?"

"I don't know, a porcelain one I guess. Quit asking me so many fucking questions."

"Okay, okay. Why get angry with me, Grant? I was only asking."

"Now chaps, I don't need to tell you how precarious things are right now." Sir John Maffey's voice rang through the mess. "I have heard the rumours, too, chaps, you know, um, that new arrivals are simply being escorted to the border – but they are far from substantiated." The diplomat had been saying the same thing for months; his credibility was definitely suffering. "For as long as the Emergency lasts, it is imperative that we respect Irish neutrality," he continued, in the face of an increasingly skeptical audience. "That's all I'm going to say about escaping. The RAF may want you to escape but I don't, particularly if the method employed offends the spirit of the oath."

Maffey fumbled awkwardly with his pipe, and cleaned his glasses waiting for the next question.

"What's this about a new camp, sir?" asked Covington.

"Yes, well, chaps, there is talk of moving you fellows twenty-five miles north of Dublin, to Gormanstown, I believe it's called, but it might not come to pass."

"Yes, but what kind of camp, sir? I've heard it's a more conventional sort of place?"

"No, I don't think so, Covie – same sort of place as this. Where did you get that idea?"

Maybe Fleming was lying; that was Covington's guess. Still, he could see the reason for it.

"Keep your spirits up, chaps," said Maffey, shifting his weight nervously from foot to foot, thinking perhaps it was Fleming, the new Canadian, who was causing all these difficulties. "Stiff upper lip, and all that. The Irish will release you any day now, it's just a matter of time."

"Will there be daily parole at Gormanstown, sir?"

The diplomat took his glasses off and looked at Covington. "Well of course, Covie. Where did you hear otherwise? Though they have been talking about splitting you chaps up, to be perfectly honest," he added. "Two buses, one to the new camp and one to the border – that sort of thing."

"What sort of thing?" snapped Covington.

This time Maffey ignored him.

"And if we escape in the meantime ..." asked Remy.

"Well, as I've said, I wouldn't advise any escape plans for the present," replied the British representative, concluding his remarks in order to dash off for cocktails with the German ambassador, or so many of them believed. "Keep quiet. That way you'll be released. At least, that's my hunch."

"Bang on, Sir John," said Ward, the former commanding officer, applauding enthusiastically from the front. "We'll do our best."

"Oh no, we won't," murmured the current one, leaning against the wall in the back.

On 30 July, Irish Derby day, Grant Fleming went to Dublin. Calder, Keefer, and Remy were with him at Jammet's, seated at the table farthest from the door. It was three o'clock. Outside, the rain was falling hard; inside, the cloth chandeliers, made from cheap burgundy swatches, hung shabbily down on the tables with spots of light piercing their seams. There was a stench of smoke and bodies left over from lunch, or maybe from dinner the previous day. No one knew for sure since no one ever noticed the smell at night. Not with all the rosemary and garlic.

There were other things that went unnoticed. The one and only window in the place had been sealed shut years ago, and the grime and soot was thick as mud. In the daytime the place looked awful.

Albert, the headwaiter, was hiding in the kitchen. By now the little man hated Maurice Remy so much that he'd just sit there, steaming, alongside the coq-au-vin. No matter what Louis Jammet said. "Let the little orphan boy starve to death," he muttered to the worn linoleum.

The man from Tipperary was expected any moment. In the meantime Fleming had finally come to grips with the fact that they knew nothing of this man from Tipperary, and that people might get killed. All that Fleming knew was what May was prepared to tell him – he was from Ballingarry, a small town in the eastern half of the county near Kilkenny. He was old enough to have been imprisoned during the Easter Rebellion – Fleming thought that she had said at Spike Harbour or Coal Harbour but he wasn't sure – and he had been involved in several business enterprises after that, none of which had been successful. He and his men had promised to come into camp with guns blazing. When Fleming jokingly suggested to May that they shoot a few guards in the process, she didn't laugh. So although the notion of using England's worst enemies to free them had appealed to him at first, it was now giving rise to second thoughts.

As Fleming looked up he saw May walking towards them in a dark green cotton dress cut just above the breasts, looking as radiant as ever. She bent over the Canadian and planted a full kiss on his lips. Pulling up a chair, she then sat down.

"I've set it all up," she said. "He'll be coming in here in a few minutes. Now where's the money, lover boy?" Her eyes gleamed beneath the sleazy cloth chandelier. Fleming, suddenly feeling dirty himself, reached into his pocket and pulled out three crisp one hundred pound notes.

"Where's the rest?" she asked.

"You'll get the rest when the man shows up tomorrow," Fleming replied.

With a frown May tucked the three bills neatly into her palm, before walking over to another table, table four, where she placed the bills under a salad plate.

"I have a bad feeling about this, mon capitaine," whispered Maurice, watching the Irish woman. "Why are we trusting this woman? And what if this maniac comes crashing into the camp and kills a few of us – Covie or me or you?"

"What choice do we have, Maurice?"

"There must be another way."

Before long, a pale, tall, thin man entered the restaurant, wearing a grey trench coat, boots, rumpled tweed pants, and a hat he hung on a peg by the door. He cursed the rain, then removed his coat, fumbling for the same peg until he realised that it had been covered by his hat. He removed his hat, dropped it on the floor, then dropped his coat, before finally picking them both up and squishing them down on the peg. He turned to look at May who directed him with her eyes to table four. The man from Tipperary was unshaven. He had dark circles beneath his eyes and wore a navy blue flannel shirt that covered a larger stomach than his pale thin face deserved.

"So this guy is going to save us," whispered Keefer.

"Ireland's in more trouble than I thought," added Jack.

The man from Tipperary walked over to table four, lifted the salad plate, and quickly pocketed the bills. He turned, glancing at the spot where May had been before she had retreated to the kitchen.

"Bet we'll never see that money again," said Keefer.

"No, I guess not," replied Fleming, as the smell of stale whiskey blew across the room. What a shame, he thought. There was something awfully appealing about getting sprung by the IRA on the same day as the Irish Derby.

Observer Calder took bomber jaunts to Rotterdam, Bremen, Emden, without mishap. Then empty gas tanks forced a parachute jump into internment in neutral Eire.

Curragh internment camp. "The last escape attempt got nowhere. Irishmen know what it is to be locked up, know all the tricks."

I Flew Into Trouble

By FLYING OFFICER JACK CALDER, R.C.A.F.

Night raids on Naziland, high jinks at station H.Q., death-dodging in flak-torn skies—Canadian now interned in Eire tells of life in the bomber command

A HAND fell on my shoulder and I turned away for a moment from the job of destroying maps, documents and instruments at the navigation desk. Keefer stood at my elbow, still carrying the hatchet with which he had chopped the rear gunner out of his jammed turret.

"I guess this means a long rest for us," he shouted above the noise of the big bomber's motors. "I never thought we would end up in Ireland."

"Maybe we'll get out of it yet, Bob," I ventured, tongue in cheek. "We've been in tougher spots."

Keefer, captain of the aircraft, went forward and took over control from the second pilot. The wireless operator fired off another Verey signal cartridge, hoping to attract attention from some airdrome. I finished clearing my desk and went up beside Keefer to peer into the darkness.

"I always wanted to make a parachute jump anyway," he said over the phone.

"But not under these circumstances," I replied. "It's a long walk home."

For four months the two of us had been flying, rooming and eating every meal together. In those four months, attached to a Royal Air Force squadron in England, we had attacked shipping and inland targets from Brest to Bremen, from Boulogne to Berlin. Now we were running out of fuel over the west coast of neutral Eire on a chilly morning. Seeing that we couldn't reach Northern Ireland or England, Keefer had picked out a hole in the thick layers of icing cloud and we were circling above it, so that when we jumped we would be sure of coming down on land.

"Check the petrol again, will you?" he asked.

"There's enough for about five more minutes."

"Okay," he ordered. "Line up the crew. You and I will jump as close together as possible. I'll yell at you in the air and we'll try to get out of the country together."

I shook hands with the four sergeants as I herded them to the escape hatch. The rear gunner, who was on his first operation, was bleeding a little about the face. He said he was all right.

The starboard engine cut out. The front gunner kicked open the escape hatch on the captain's order and dropped through. The second pilot, rear gunner and wireless operator followed.

Jumping was easy. As my feet went out, the slipstream caught them and I was speeded through by the rush of air. I pulled the rip cord; in a moment my head was jerked back as the chute opened and I felt as if I were being sawn through. The sensation ended quickly and the first thing I noticed was the desperate quiet after eight hours of listening to the buzz of the engines and the crackling of the telephone.

I shouted for Keefer and got no reply. Then I heard the low whirr of our aircraft—old "C for Charlie"—gliding to her finish. Bob had planned to head her out to sea, but now she seemed to be coming around. I learned later that Keefer had headed her for the sea, started to jump and then had seen she was turning inland. He returned to the controls, pointed the machine for the Atlantic and jumped. The aircraft came around again and finally broke her back in a pasture.

Down In A Bog

FOR A long time I seemed not to be falling at all, just swaying a little in the breeze. I turned around to make sure that I wouldn't be carried out to sea and I inflated my "Mae West" just in case. Suddenly I realized the ground was near and I relaxed for the impact. Nothing happened. I bent my knees. My feet hit the ground and my knees hit my chin.

I got up, released my chute and tucked the folds under my arm. In the darkness I could see absolutely nothing but the stars and a couple of lights on the coast. I decided to start north and try to find Keefer. I stepped out smartly and immediately went to my hips in water. I retraced my steps and headed south; this time I sank to my knees in ooze and goo. Attempts west and east brought the same results.

It dawned on me that I was in a bog. It really wasn't a bad bog as Irish bogs go, but I was stuck there until dawn. I sat down on my little dry spot and ate a bit c chocolate, wondering all the while how the res of the crew were faring. My head ached.

As the first light of dawn flooded the bog half-hour later, I discerned in the distance what thought to be a great wall. The light grew and, jus about twenty yards from me, I picked out a ribbo of road. If I could have got to that in the darkness I might have stumbled to some sort of hiding place

Now I hid my parachute, waded to the roadwa and fairly ran along it toward the wall, for ther wasn't a sign of other cover. As I got quite close t the wall, however, I saw that it was made o piled-up blocks of peat, the winter's fuel supply fo hundreds of Irish families.

I hurried past and came upon a gate bearing sign: "Keep Gate Shut." The thought came t me: "Well, we speak the same language. Mayb they'll listen to an argument if they catch me."

Deciding to head for the coast, I tore the badge and insignia from my flying dress as I walked Around me were the tiny white cottages of th turf-cutters, each with a noisy dog. Smoke curle above the thatched roofs.

The sun started to rise and, in desperation, looked for a hiding place. A small copse close t a cottage seemed the most likely, and furtivel I climbed the low stone wall to go to it. As I did so a child came from the house.

"O, momma," she cried. "Lookit the man!"

The family came out and I hurried awa wondering whether I was lucky enough to resembl an Irish vagrant.

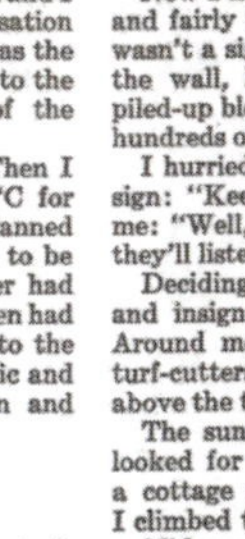

Jack Calder, author of this article, before the war a Canadian Press news writer.

The countryside was astir b now. Cattle drovers and donke carts were on the road. Alon the way I said good mornin cheerfully to everyone. Da O'Fagan said good morning fro his pub door and the children san good morning from the fence posts

"Not bad," I thought, wonderin if my red hair made me look lik a native. "It can't be more than hundred miles to the border, either."

When I had walked eight or te miles I sighted a railroad track an then a patch of woods overlookin it, near a coastal village. It looke like an ideal spot to wait for

Continued on page 25

Above and opposite page: *Maclean's*, 15 August 1942

I Flew Into Trouble

Continued from page 11

freight train, and I started toward it.

"Hello," I was accosted from the roadside. "Where are ye goin' and where are ye after comin' from?"

"I'm from the south and I'm going north," I told the Irish policeman, just as if I were addressing a traffic cop back home.

"Well, them as goes north always comes into our guard barracks for a nice cup o' tay," he smiled. It was a real, nice, top-o'-the-mornin' smile, but I could feel him frisking me with his eyes for firearms. There was nothing for it but to go along.

"We heared ye wuz comin'," he said. "Five others like ye will be along in a twinkle." And then, wistfully, "Sure an' ye're all just lads, too."

In the police barracks I was fed bacon and eggs and Irish whisky, which goes down like fire and hits bottom like a sledge hammer. Keefer and the English sergeants came along soon. Bob was limping because, in landing, he had revived a knee injury which he suffered when he played with McGill University's football team.

"I had to go to a cottage," he reported. "The people said they would smuggle me into the north. So, while I was eating their bacon and eggs, they got the cops. Hey, look! Don't touch that Irish whisky. They gave me a shot and I nearly went through the floor."

The stories of the sergeants were similar to our own. They were in good shape.

The police and militia, arriving by the dozen, questioned us from every angle. Bob and I lied variously (and obviously) that we had brought a hundred parachutists and dropped them all over Ireland; that we had been flying alone and didn't know who the sergeants were; and that we didn't think we should be kept in a police station when our mothers and fathers came from Ireland and had been greatly interested in the Gaelic revival in America and the police strike in Boston. Keefer's name was O'Keefer and mine Fitz-Calder.

(I have learned since to swear that I am descended from the ancient Kings of Ireland and came "home" deliberately.)

When we had found out that we were at Quilty, in county Clare, we got permission to go outside and lie in the sun, for it was chilly in the barracks. We hoped, of course, to be able to get away, but the numbers of troops outside dashed those hopes quickly.

Bombs On Rotterdam

A BRIGHT sun shone on the sea, which lay calm and blue in the rock-bound inlet a mile away. All about us the land was green and in the background lay purple mountains.

While scores of Irish country folk stood gaping at us, Keefer fell asleep and I fell to thinking. I thought of the jobs we'd done and the fun we'd had together. Though I had been awake for more than twenty-four hours, I couldn't sleep. The letdown had been too sudden.

I remembered the tingling excitements of my first trip. It was to Rotterdam, where I released my first load of big bombs in the dock area. I was flying that night with Slapsy Maxie, a New Zealander who had been through the business in France and had flown the Atlantic twice. While he took avoiding action, he explained the various kinds of tracer shells that were being fired at us and the systems of searchlight fingers that were trying to pick us out, as if he were giving a lecture back in the crew room.

When we went to Frankfurt a couple of nights later, we were bounced around a bit by heavy anti-aircraft fire near Brussels; Maxie laughed while he worked the throttles to desynchronize the motors. On we went to find a terrific wall of fire being thrown up in one sector at the target area.

"There's something funny here," said Maxie. "I think they want us to believe the target is where they're shooting from. But do you see that bend in the river down there? That's what we're looking for."

When I dropped a flare, Maxie's opinion was confirmed. We were able to let the load go quickly and go back to England, where we were diverted to another airdrome because our own was under fog.

I did Brest in daylight with Maxie. It seems to have been the most exciting episode of my life, for our formation shot down three Messerschmitt 109's in our run-up to the target. Maxie got the Distinguished Flying Cross and the gunnery leader the Distinguished Flying Medal for that one.

Many of the jaunts with Keefer were nearly as successful. We couldn't have hit the target more squarely than we did on our first together. We just glided into Boulogne with a full moon ahead, released our stick across the docks and glided out again before Jerry knew we were there. The Air Ministry has recorded that the fires were burning the next day.

The things that stood out most prominently, though—as I lay on the green, green Irish lawn and gazed into the blue, blue Irish sky—were the things that had gone wrong and become right again, the things we had gone through and lived to tell about.

8 A HANDY CORKSCREW

When the *Maclean's* piece finally came out on 15 August 1942, providing Colonel McNally with clear evidence that the former reporter was indeed making the best of a difficult situation (or, as Jack put it at the close of his article, enjoying the things that had gone wrong and then right again), Fleming and Keefer were out on the camp links, making the best of theirs. Having ultimately rejected the notion of using England's worst enemy as their ally, each was racking his brain for some other solution to the vexing problem of the guards, and the wire. Standing in the rain (as Girdlestone once noted, it rained so much during their internment that the angels must have had a bathroom above them) or sitting in one of the many feed shelters dotting the fairways, they thought of every possibility. The hope was that something would come to them, knowing that by even talking about escaping in such circumstances they were breaching their parole, but no longer caring. Not when the RAF no longer cared about them.

And then, all of a sudden, the answer came, if not like a ribbon of flak piercing the night air then at least from an unexpected source. Keefer, who had been describing to Fleming his summers spent on the locks of the Erie canal in the Hawartha Lake district in southern Ontario with his brother Eddy, suddenly remembered an old farmyard gate that the two used to swing on. This was back on a neighbour's cattle farm, where his mother had reserved a family vegetable plot. The gate was mounted on gudgeon pins and, like the nearby canal locks, it would either slide open or lift off its mounts. One day he and Eddy accidentally dislodged it and the cattle stormed in, which resulted in a period of internment then, too. Whether it was Fleming or he who then remembered the double gate on the southwest side of the camp really doesn't matter.

Of course, part of the problem was that no one had ever been over to that corner before. They had avoided it during the 9 February attempt, and in all previous attempts, for that's where the Irish Army's main rifle range was – not an area they were anxious to visit with their backs turned, running down the road. However the two knew that there was an entrance there, because people had seen trucks coming from that direction, delivering materials used in the construction of the new buildings. And when Keefer and Fleming played the front nine a second time that afternoon and made a short side trip into the gorse (nothing unusual about that) they were delighted to discover that the two gates were similar to those Keefer had described.

What did they have to lose?

They told the other two that night and everyone agreed. They'd golf tomorrow, rain or shine, and take a closer look.

"Are those the gates, Jack?" whispered Fleming, as he, Keefer, and Girdlestone joined them the next day on all fours, crawling on their bellies through the gorse, bullets whizzing high above their heads. (This was after Keefer's second shot, a screaming three wood, had rocketed off the fairway at right angles, landing 200 yards out of bounds. No penalty stroke, that was the deal.) On closer examination the gates were about six feet wide and twelve feet high,

The 9th tee. Though from my observations it would have been rather crowded with four of them in there!

lashed together by a padlock and sturdy chain, with wire lattice and metal reinforcement. The gudgeon pins, like the ones of his youth, were L-shaped brackets that swung open to the inside, hanging the weight of the gates on nothing more complicated than four large spikes driven horizontally into the post and bent upwards to ninety degrees. They might be able to hoist the whole thing off with two men on each gate, lifting up and out simultaneously. It was at least worth a try.

The next thing that had to be decided was who else to tell. All four – Fleming, Keefer, Girdlestone, and Calder – suspected that the guards must have known about the 9 February attempt in advance, so quick were they to react with reinforcements. With eleven officers and twice that number of NCOs it wouldn't have been surprising – a joke, an offhanded comment, and suddenly the jig is up. So Fleming decided that the matter would remain a secret until the evening of the escape, which would be closer to midnight this time. No one was to know, especially Ward.

As luck would have it, the Irish Derby was scheduled for that weekend. The Irish Derby, like the Harrier's Ball on New Year's Eve, was one of those functions that just couldn't be missed, an engagement that required the attendance of not only the Horse Protestants, but the wealthy merchants of Dublin, professionals, a few politicians, some retired Irish Army officers, English cottage vacationers, international horse breeders – in sum, everyone who was anyone on the Pale. This provided them with the opportunity they were looking for – getting everyone together in the same place and discretely seeing who would help.

The Derby was always an impressive event, with the Grandstand at the Curragh's main track filled to the brim, as it was for the Oaks, the St Leger classic and the 1,000 and 2,000 Guineas, the other races that year. Adjoining fields were replete with horse carriages and models T's, with servants, nannies, and stable boys in tow, and, of course, British internees, the county's most famous guests, mingling alongside. There was the runup to the big race as well, dinner with the O'Sullivan's on the Thursday night, a champagne breakfast with George and Margaret Mullans on Friday morning, dinner that evening with the Dalys at Russborough, back to camp for some more banter with the guards, and then back out to the track on the

The Grandstand at the Curragh, County Kildare. Courtesy Bill Gibson

day of the race for some claret and Golden Plover with the girls in Major Mitchell's private box in the morning, and a cocktail or two with Sir Haldane Porter in his Guinness box that afternoon. Sir Haldane was a big supporter of theirs, having hosted Covington, Keefer, and Girdlestone on a tour of his brewery the previous week. They had even met his science director, Dr Fleury, who, over pints, had told them all about how the water was taken from the Liffey and then used for the bottled stout that was exported, while a secret water source to the east in Kildare was used for the draft. No self-respecting Irishman would ever drink from the Liffey. Both Dr Fleury and Sir Haldane were impressed – not only by the attentiveness, but by the consumption rates at the officers' bar, which they had heard about from Sir John Maffey.

After all was said and done Keefer decided that he would impose upon two people, Aubrey Brabizon, one of the better-known jockeys in Kildare who was also the head trainer at the Aga Khan stud farm, who had offered to put them up in a stable for a few days (what the Aga Khan didn't know wouldn't hurt him) and Major O'Sullivan, Sheila and Maureen's father, who a few days earlier had offered to drive them to the North. "I know you boys are out on parole and aren't supposed to be talking about his," said the major, as the three sat in his drawing room overlooking Dublin Bay. "But if you ever do get out you can count on me. I'm not about to break any laws of the land but … enough said."

Keefer, feeling very much like the cat who had swallowed the canary, just nodded, secrecy being the order of the day.

With Brabizon, though, he was a little more direct, requesting and receiving directions to a stable and a loft above that no one would suspect.

On the morning of Monday, 17 August 1942, all four awoke early. The clouds hung thick over the plain and Fitzpatrick would be off duty that day, another good sign. It was agreed then: the escape would proceed that night.

While no one had to build a ladder this time, they still had to find some metal bars somewhere, strong enough to lift the two gates. Once more they turned to the NCO's bath house, dismantling two of the shower curtains (tossing an NCO out in the process) and hiding the bars in the bicycle shed later that afternoon. That way, the guards who inspected their quarters every six hours would be none the wiser.

Fleming, with Keefer, Calder, and Girdlestone at his side, then made the announcement first in the officers' bar, and later in NCO's mess. The break was set for 23:00 hours, sharp. Whoever wanted to join them was welcome.

"Listen up," said Fleming, as all eyes and ears turned his way. "Some of us are planning a break tonight, and we need six volunteers." Hands shot up. "The plan is to break through the wire over by the rifle range." A few worried faces appeared, though no one lowered their hands. "Good. There are two lorry gates over there, locked, but we think that enough upward pressure might spring them free from their hinges." Fleming looked over at Keefer. "Now, frankly, I don't know if this is going to work, but the only way to find out is to try. Here's the plan."

Fleming then went on to explain that two officers had already been chosen for the first two jobs. Calder and Wolfe would proceed to the police hut on their bicycles as if signing out for parole. They would then overpower the guard and jam one of their bikes into the gate, thereby allowing the balance of them (seventeen, plus anyone who arrived at the last moment) access into the outer compound without scaling the inner wire that had proved so difficult last time. In order to get everyone through the police hut, they'd have to move fast.

Next, in order to cut off the reinforcements from the duty hut in the outer compound, they needed someone to jam that lock. Covington suggested a corkscrew from the officers' bar – brilliant.

And finally, among the NCOs they needed four men strong enough to lift the gates, two on either side using curtain rods as levers. Sgts Newport and Harkell, two beefy Brits, were chosen for one rod, and Brady and Keefer for the other. Of course, Keefer wasn't an NCO, but no one else was strong enough. After that, they'd still have to scale a five-foot wire barrier on the outside and worry about the soldiers taking aim at them from the rifle range next door. But everyone agreed that shouldn't be a problem, not at that hour. They dashed off to prepare their escape kits.

As zero hour approached, Fleming stood at the corner of his hut, with Keefer on his right and Girdlestone on his left. Everyone was lined up ready to go, with darkened faces and bits of sliced mattresses once more strapped to their arms and legs to protect them from the wire. Fleming paused one last time to think about what would have happened had they proceeded with their IRA contact, the man from Tipperary. He shuddered.

At 23:00 hours sharp, Bud Wolfe appeared from the cycle shed, pedaling his way towards the police hut. Jack fell in behind. When the two arrived, Wolfe greeted the policeman there, and bent down to adjust his bicycle clips. "All right Jack, you go ahead," he said, "I'll just be a minute."

As Jack passed through the gate ahead of him, the American quickly grabbed the startled policeman by the neck and pinned him in a headlock. Jack then jammed his front wheel in the gate. A second later the first six internees arrived on cue. They squeezed quickly through the opening, as Calder and Wolfe continued grappling with the guard. Covington, who was first out, dashed over to the duty hut and deftly slipped the corkscrew into the lock – mission accomplished. Keefer, Brady, Harknell, and Newport then moved across the compound towards the locked gates, curtain rods in hand. Still no alarm had sounded, and all seemed well. Half of them were out now, and the police hut was quiet – apart from a muffled cry or two.

None of the turnkeys patrolling the outer fence had noticed a thing, either. Even the duty hut was quiet, blissfully ignorant of the surprise that awaited them.

Keefer, Brady, and the English sergeants prepared themselves for their lift. Wrapping his gloves firmly around one end of the curtain rod, Keefer looked around. Would it work? It certainly had so far – none of the reinforcements were anywhere to be seen! Fleming and Girdlestone were there too, with the former waiting to give the command. Suddenly the guard patrolling the outer gate seemed to notice something. He took three or four steps towards them in the darkness, peering through the wire directly opposite them. Keefer looked up and caught Fleming's eye: would this finally be the moment that one of them got shot? The guard opened his leather holster and drew his gun, just as the four maneuvered into place for the lift. The Irishman was no further than six feet away now, pointing his weapon directly at Keefer. Had he seen him? Would they have time before he began to fire? What were those zany British internees up to anyway?

"GO!" shouted Fleming suddenly, as the four internees lifted in unison for all they were worth. After a second or two when it appeared that no number of men could move it, the double gate lifted from its hinges and fell over – directly onto the guard, knocking the gun from his hand and pinning the hapless sap to the ground.

Unbelievable! It had worked!

The four then took off like a shot toward the golf course, with Fleming and Girdlestone, the two Poles, and Fowler, falling in behind. A quick look over his shoulder told Keefer all he needed to know. Nine of them had made it out, thanks to his plan and to Covington's handy corkscrew. The others, including Jack, were nowhere to be seen.

Within minutes they had broken into three groups, with Girdlestone heading north, the NCOs heading west, and the two Canadian officers continuing south, past the eleventh tee and Donnolly's Hollow. With the sounds of screeching whistles and barking dogs behind them, Keefer and Fleming were soon crossing roads and jumping fences until they had doubled back and were heading north, toward the Aga Khan stables where Brabizon, the jockey, had offered them shelter. It was then that they encountered their next obstacle. At first they thought it was a motorcycle or maybe a lorrie charging across the field at them. After diving for safety, however, they discovered it wasn't a truck at all. It was a bull – appropriate – since no one back in their squadrons was going to believe this. After a brief rest, they continued at a slower pace, carefully avoiding lights, buildings, or any sign of movement, until some two hours later when they reached the Aga Khan's.

At that time the Aga Khan stud farm had three or four tracks and several hundred stables. At a row of stables in the back of the estate, they paused to drink heavily from a water faucet before entering the last stall in the row furthest from the road. Keefer climbed the ladder along the right wall, which took them to a small landing above. There they found a trap door. Inside was a loft with a peak 3 1/2 feet high, which meant that once inside they'd have to crawl. The plan, which had been hastily drawn up over the last day or so, called for two or three days there, with rations of turnip and dried meat. At night they could stretch their legs by climbing down to the stables below, but during the day, when the grooms and stable boys were down there, they would have to lie perfectly still. It was 4 am before they finally got to sleep. Still, things could have been worse, they agreed. They could have been back in camp.

On the morning of the fourth day no one had arrived to take them to the North. Ostensibly that was because Major O'Sullivan had screwed up, though in truth neither had firmed up the details with him, so it might not have been his fault at all. By then they were growing impatient, relying on distractions such as eating their small meals as slowly as possible or playing a variety of Scrabble where one person writes a word and discloses only the number of letters, while the other guesses what the word is, hoping that his friend hadn't misspelled it. That worked fine until they got onto Irish place names. They decided that, come nightfall, they would set out for Belfast on their own.

Once darkness had fallen, and with Fleming's map as their guide, they headed north, walking throughout the night. It was impossible not to notice the number of soldiers out on the roads, the lights, the whistles, and the dogs, a full three and a half days after the break. Maybe that meant that the whole camp had escaped – all the better for them. Nearing nightfall, they stopped for a shave in a cool stream. When Keefer enquired as to why the sudden concern with their appearance, he was told that they would now be resting at the Kelly's estate in Sallins north of Naas, which was a backup plan that Fleming had arranged on his own. You don't get to be a war hero for nothing, thought Keefer, though he was a little miffed that Fleming hadn't consulted him. Still, Mr Kelly, a retired businessman and father of Mary, Grant's girl friend, seemed like a nice enough man. They could do worse.

At dawn on the morning of the following day, Monday, 24 August, they arrived at Sallins. Tossing a few pebbles at her window, Fleming woke Mary, who directed them in a whisper to a nearby barn. There they waited, until she and her parents showed up with hot food, tea, and cookies. As those were being consumed, they learned that there had been a landing at Dieppe a few days earlier with Canadian casualties (little did they know) and that only three internees were still at large. Mr Kelly even produced a clipping from a local newspaper to prove it. Rumour had it that the third was an English sergeant, Vic Newport, who had already made it to the North. The last of the six to be captured had been the New Zealander, Girdlestone, who had been nabbed in a biscuit factory in Dublin after the garda had chased him there in the company of two members of the notorious Escape Club, Dr Tom Wilson and Dicky Ruttledge. Keefer and Fleming knew about the Escape Club,[35] but avoided it. A pro-British group bent on resurrecting Protestant glory in the South, they tended to attract attention. Finally, Mr Kelly mentioned one bit of good news: for some reason all of the soldiers in

NINE BRITISH INTERNEES

ESCAPE IN EIRE.

THREE RETAKEN.

It was officially stated in Dublin to-day that nine British internees escaped from the internment camp at the Curragh at 11 o'clock on Monday evening.

Three have been recaptured, but the others are still at liberty.

The clipping that Mr Kelly produced, Monday, 24 August 1942.

ANOTHER INTERNEE RECAPTURED

Six of the British internees who escaped from the Curragh on Monday week have now been recaptured. Three are still at large.

It was announced by the Government Information Bureau, on behalf of the Department of Defence, yesterday, that the sixth internee to be recaptured was taken into custody during the week-end.

A clipping from the same local paper. In fact, three of the nine, Sgt Brady and the two Poles, Baranowski and Karniewski, had already been captured by then, Keefer's fourth day on the lam.

35 The archives contain numerous references to the Escape Club, an organization set up by friendly Horse Protestants and retired army officers to help the internees to the North if and when they ever cleared the wire. One reference details the arrest and prosecution of Dr Tom Wilson, a well-known Dublin physician, for his role in aiding Girdlestone's unsuccessful escape. Wilson was eventually convicted and sentenced to nine months in jail, a term that was suspended at the last moment after the British Representative's office agreed to pay £300, a high fine back then but a steal of a deal compared to what Begg was prepared to pay the IRA.

the fields and at the roadblocks had packed up and left that afternoon. It would be weeks before that odd fact was explained.

Still, just to be on the safe side, it was decided the following morning that the two Canadians would be well served by wearing disguises. By the time night came, therefore, as Grant Fleming kissed his girlfriend goodbye, the two had become cattle drovers. Mary's father cycled north with them to Kilcock, where another friendly face greeted them at dawn. That was Major Aylmer, who lived on the Kildare border at Courtown. Having followed the Liffey through most of the county, OldConnel, Barrettstown, Carragh, Clane, and RathCoffey – they would now be heading into less familiar territory. Still, their general approach had been right so far: Sleep during the day, avoid the more traveled roads at night, and be eternally grateful for the assistance and support coming at them from every direction – knowing how diligent Commandant MacKay and his men could be. Interestingly, of all the stately Georgian homes in which Keefer was entertained during his Irish internment, none surpassed the Aylmer estate. It seemed fitting, as his final days on the Emerald Isle were approaching.

The last link in the chain was to be a retired Royal Navy officer known only as the Captain.[36] After a dinner of cold chicken with Major Aylmer, Fleming and Keefer were sent off to meet this man with little more than a bicycle each and a map. Apparently he lived in Monaghan to the north.

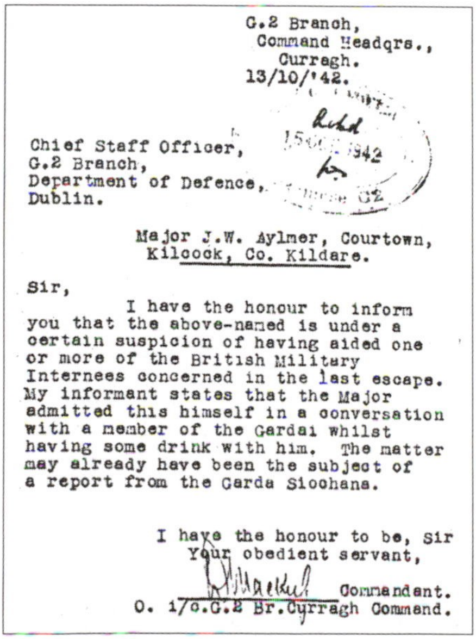

G.2 Branch,
Command Headqrs.,
Curragh.
13/10/'42.

Chief Staff Officer,
G.2 Branch,
Department of Defence,
Dublin.

Major J.W. Aylmer, Courtown,
Kilcock, Co. Kildare.

Sir,

I have the honour to inform you that the above-named is under a certain suspicion of having aided one or more of the British Military Internees concerned in the last escape. My informant states that the Major admitted this himself in a conversation with a member of the Gardai whilst having some drink with him. The matter may already have been the subject of a report from the Garda Siochana.

I have the honour to be, Sir
Your obedient servant,
Commandant.
O. i/c.G.2 Br.Curragh Command.

From the head lackey himself. Fortunately they only discovered this major's complicity after the fact. Courtesy Irish Military Archives

36 This man was never reliably identified, by either Keefer or the archives, beyond his rank as a retired naval captain.

They began by following a tributary of the main road north, with no lights, and their cattle drover caps pulled down low over their eyes. After following the road through Garadice, Laracor, Trim, Navan, and Slane, they finally arrived at Ardee. They turned off the main road sometime after 3 AM, following the well-marked map to the captain's cattle farm. Surprisingly, they found that with little difficulty. There they were introduced to a large, friendly, bearded old sailor who poured them several stiff glasses of Irish whiskey. He didn't think much of their disguises and advised them that as of the following morning they would no longer be cattle drovers. Instead they would be his Australian cousins. He then showed his Australian cousins to separate bedrooms in a large farmhouse.

The following morning, Wednesday, 26 August, 1942, Fleming and Keefer, having boned up on their Australian slang, were introduced to the captain's servants. Knowing from their experience on the Curragh that one could never be too careful around such people, the two kept conversation to a minimum, other than the occasional "g'day", or "owyagowin' mate?" After breakfast the captain announced that it was harvest time and that his two cousins from Australia were going to get some exercise. He then introduced them to his two Irish farm hands. The fields were dotted with haystacks, and before long all four were tossing bales into the back of a wagon with their pitchforks, talking about women and Australian beer, and so on. In fact it was one of those rare occasions in Ireland where there wasn't a cloud in the sky all day long. The work was strenuous but enjoyable, and by the afternoon the cousins had their shirts off, basking in the sun and saying things like "jeez, its figginot today," "'where's the barbie,'" and, if anyone complained, "well, it's better than a poke in an eye with a burnt stick, don't ye think mate?"

"Bobby, I'm thrashed," said Fleming, quietly, as they sat beneath an oak tree cooling themselves at the end of it all, finally finding some privacy. The farm hands had headed home and the captain had gone back to fill his mickey. "It's been a long time since I've had to work like this."

"I know," agreed Keefer. "Nice though." But then he suddenly felt angry, almost in spite of himself, looking around at the scene before them, the deep blue of the sky, the many shades of green, the rolling hills. Even the captain's jolly appearance, and the smell of stale whisky, couldn't shake the feeling.

What was it? That he missed Jack? Sure, that was part of it – although they were still miles from freedom. That he hadn't met someone like his two friends had, Ann, thinking of Jack, or Mary, thinking of Grant; having to bend down and start adjusting his bloody bicycle chain so that his commanding officer could kiss his Irish colleen goodbye. How humiliating!

No, that had been his choice. He had, after all, enjoyed a few nights with Josephine at Osberstown House in the privacy of her bar after closing. At least he'd have that to remember.

There was something else, he realized, something altogether different that was bothering him, something that he hadn't expected at all. It was something that had begun long before that, maybe with their landing in Clare, meeting the lads from the local defence force. Or maybe Covie and the others at the officers bar, Susan and Sheila, or even Shaw, who he couldn't stand, who made him get dressed in those stupid jodhpurs. Or maybe New Year's Eve and the German Oberlieutnent Neymeyr, when he suddenly caught himself wishing the man well. Or maybe all of the good times that summer with Jack when he finally corrected his slice, when he had learned that cardinal rule of golf: inside out and let the club do the work.

And then, finally, as he sat there on the top of the hill looking down on the scene below, it came to him. For the first time he realized how he really felt about the place. About Ireland.

That he hated it. Goddamn it. He hated it. The religion, the politics, the priests, the pubs, the screeching pennywhistles, the silly accents, the horses, the women, and now this: like a guest who had spent his welcome, he was just expected to get up and go, no goodbyes, no time for regrets, just go. How dare Ireland do that to them. How dare the bastards not understand.

Keefer and Fleming finally reached northern Ireland early Thursday morning, 27 August, 1942. Like their arrival in the south, their arrival in the north was unexpected, and more than a little comic. Led by the captain to a point along the border immediately south of Roslea, they followed a path two miles through the woods until they reached a red pillar or mailbox, with its distinctive coat of arms. Aided by the light of a match, they determined that they had crossed the border, for the boxes in the Republic, like everything else, are green. They then set out to find a police station, soon com-

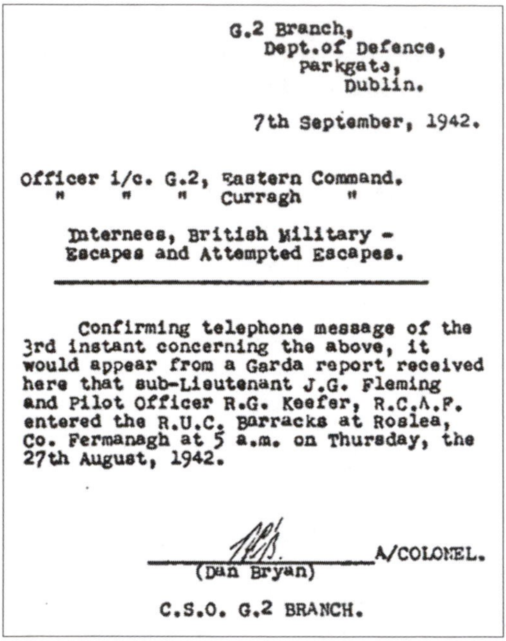

G.2 Branch,
Dept.of Defence,
Parkgate,
Dublin.

7th September, 1942.

Officer i/c. G.2, Eastern Command.
" " " Curragh "

Internees, British Military -
Escapes and Attempted Escapes.

Confirming telephone message of the 3rd instant concerning the above, it would appear from a Garda report received here that sub-Lieutenant J.G. Fleming and Pilot Officer R.G. Keefer, R.C.A.F. entered the R.U.C. Barracks at Roslea, Co. Fermanagh at 5 a.m. on Thursday, the 27th August, 1942.

A/COLONEL.
(Dan Bryan)

C.S.O. G.2 BRANCH.

Courtesy Irish Military Archives

ing upon a building on the main street, once more with a bronze plaque. This one read RUC – the Royal Ulster Constabulary. That sounded promising. They banged on the door in their long cattle drover's frocks with their caps pulled down over their eyes, until finally it opened with a crack. They quickly identified themselves as RAF internees who had just escaped from the Curragh, flashing the identity disks, or dog tags, around their necks. Just as quickly, the door slammed and a voice inside ordered them to go back to where they had come from.[37] A moment later, however, the door opened a second time and three policemen appeared, with guns. Real ones.

"Sure you boys aren't IRA?" they asked, pointing their guns directly at the startled Canadians who had somehow managed to avoid getting shot to that point. "We've had some Troubles lately."

They were sure.

From there they were taken inside and given a phone call. Fleming reached the CO at the RAF base in Aldergrove and soon a car was on its way. In the meantime, they sat down with the police and had a beer to celebrate.

37 Like the Wizard of Oz, he used to say, only ruder.

An hour later, a RAF wing commander appeared. He was driving a Jaguar convertible with the top down. The two tired but happy Canadians introduced themselves and jumped in the back. After a short debriefing – "Congratulations, lads, on getting out of Ireland," said the Wingco, "it's certainly a lot more commendable than getting in." – they were billeted for a day with a retired army officer in Belfast where they slept around the clock. The following morning they were flown to London in a small, four-seat German Fokker. Soon they'd be back in the racket and, for some reason, they were pleased as punch.

G.2 Branch,
Command Headqrs.,
Curragh.

29 JUL 1942

28/8/42.

Chief Staff Officer,
G.2 Branch,
Department of Defence,
Dublin.

Escape of British Internees 17/8/42.

Sir,

I have the honour to confirm telephone conversations re above and to state that at 22.50hrs. on the 17th inst. 9 British Internees escaped from internment at the Curragh. The escape was engineered as follows:-

Pilot Officer Wolfe appeared at the inner gate and signified to the P.A. on duty that he wanted to get out on parole. With him was P.O. Calder. The P.A. opened the gate and Wolfe was nearly through with his cycle when he said "excuse me I must put on my trouser clips – you go through Jack" (to Calder). Calder pushed by Wolfe and when inside the cage both held the P.A. The P.A. at the outer gate gave the alarm but by this time the cage was full of internees who brought with them pipes removed from the wash house which they used as levers on the outside gate taking it off the hinges. Seventeen internees in all took part in the attempt, eight were pushed back by the available P.As. and 9 escaped into the darkness. They were:-

F/Lieut. Fleming,
P.O. Keefer.
Sub. Lt. Girdlestone.
F/O. Baranowski.
Sergt. Brady.
Sergt. Karniewski.
Sergt. Fowler.
Sergt. Newport.
Sergt. Harkell.

A cordon was established around the Curragh, the Garda notified and a search was made for the escapees.

During the night F/O. Baranowski, Sergt Brady and Sergt. Kernieawski were taken into custody all being taken by Military and Garda in the vicinity of Kildare. Patrols were kept on and searches made up to the 22nd. inst. and P.As. sent to Dublin to assist there. On the 20th. Sergt. Herkell was taken into custody at Dolphins Barn, Dublin and Sergt. Fowler was taken into custody at Tallaght by Military, and Sub. Lieut. Girdlestone was taken into custody by P.As. in Dublin, back by the Post Office Curragh and verified that they did originate in Belfast so it would seem that Fleming succeeded in getting across the border. There is still no trace of Keefer or Newport.

I have the honour to be,
Sir,
Your obedient servant,

D Mackey Commandant.
(D.Mackey).
Officer i/c G.2 Branch,
Curragh Command.

While several reports were filed for this escape, only two are reproduced here, Mackay's report and that of Mr Kearney, the Canadian Commissioner. Courtesy Irish Military Archives

Office of the High Commissioner for Canada
Dublin
August 31, 1942

No 224
Sir,

I have the honour to refer to my telegrams No. 58 of August 21st, and No 65 of August 31st. The following are the details in connection with the successful escape from the Curragh Internment Camp Co. Kildare of F/Lt J.G. Fleming, D.F.C. and P/O R. G. Keefer:

On the evening of August 17th, P/O R. N. Wolfe, a national of the United States of America, attached to the Royal Air Force, and P/O Jack Calder of the Royal Canadian Air Force were going to a party. When Wolfe, who had his bicycle with him, reached the first gate, he stooped down to adjust his bicycle clips and said: "Alright Jack, you go ahead, I won't be a minute." (Only one officer is allowed to go through this gate at a time.) No sooner had the sentry opened the gate for Calder, than Wolfe threw his bicycle between the gate and the gate-post, thus preventing the gate from closing, and they both then over-powered the sentry. Immediately a party of about thirteen or fourteen internees rushed the first gate and with two iron bars, which they had dismantled from the shower-bath, lifted the second gate clear off its hinges. This unexpected maneuver took the sentry who was posted at the second gate by surprise, and the gate fell on top of him. The internees had then a third and outer ring of barbed wire to clear. It was not very high and nine of them managed to scramble over it with no worse damages than torn clothes and a few slight scratches. The others attempting to escape were captured before they could surmount the third ring of barbed wire, but nine, four of whom were Canadians, namely F/Lt. J.G. Fleming, P/O R. G. Keefer, Sgt. G. D. Fowler, and Sgt. C. S. Brady, managed to escape.

It so happened that the Irish Army was on maneuvers in the neighbourhood of the Curragh camp and an intensive search was immediately made. Brady was at large for about an hour. Fowler was out all night and the next day, but was apprehended in the neighbourhood of Dublin and taken to the Garda, (police station) from where he made another effort to escape but was again recaptured. Fleming and Keefer were secreted by some friendly Irish families and were assisted over the Northern boundary. The only other internee not recaptured is an English Royal Air Force Sargeant by the name of D. B. Newport. A New Zealander by the name of Sub-Lieutenant B. N. Girdlestone, after being at large five days was recaptured in Dublin. He had been hidden by a friendly Irish family who attempted to convey him to the North but their car was followed and he was taken after a lively chase.

The fact that it is now known who the Irish family that assisted Girdlestone may lead to some trouble, and they are being interrogated. A woman by the name of Roseen Bruagh, who is a daughter of Cathal Bruagh, one of the men who was executed following the Easter rebellion in 1916, hid the German parachutist who escaped from Mount Joy prison but was later recaptured; as soon as this was known she was interned. The British Representative's office have therefore some fears that some action will be taken against the family who assisted Girdlestone.

The escape was very well organized and executed. Only a few internees actually knew of the attempted escape until the last few minutes, and for this reason, several were out on parole, including F/Lt. L. J. Ward. Ward was not told about the plan as his wife who lives in England was on a visit to Ireland and the other internees thought it might give rise to suspicion if he came back to camp.

Mr Garland and I visited the internees and I had an interview with Colonel McNally, the Officer commanding the Curragh Camp. Colonel McNally considers the break a clean one, and expressed admiration for the ingenuity with which it was effected. He told me that the German internees were very critical of the Irish Army and were spreading a rumour that the sentries who were on guard over the Allied internees opened the camp gates to allow them to escape. He stated that it was difficult to explain that to anybody, let alone to the Germans, how those who escaped could break out and elude so many guards. He is very well disposed towards the Allied internees and regrets having to keep them in confinement.

Following the last escape, on August 17th, all parole for the internees was cancelled and Irish soldiers were stationed on all likely roads of escape. Ward's wife was not allowed to see him even through barbed wire. Sir John Maffey, the British Representative, made representations to the Irish authorities and he asked me (because Canadians were involved) to also interview the Irish authorities so that confinement would not last for a month as it did following the attempted escape which was made on the 9th of February last, and also to try to impress upon the Irish authorities that they had done quite well enough in recapturing six out of the nine who had escaped, and that they should therefore lessen their intensive search for the remainder.

I, accordingly, saw Mr Washe, Secretary of the Department of External Affairs. He was sympathetic but stated that it was difficult to drop any hints to the Army. However, he said that he would do what he could in the matter. My intervention, I think, may have been of some assistance. In any event, the day following my interview with Mr Walshe, parole was restored while Fleming, Keefer and Newport were still in Ireland and the extra guards on the roads were removed.

Fleming and Keefer arrived in Belfast on the 28th of August; shortly afterwards they proceeded to London. I understand they are in good health, and delighted to be able to rejoin their units.

I have the honour to be,
Sir,
Your most obedient servant,
John D. Kearney
High Commissioner for Canada in Ireland

9

THE NEW PLAN

Meanwhile, back at the Curragh, Jack Calder was not faring so well. As one of nine internees caught in the outer compound when the corkscrew from the officers bar finally popped out, Jack, like the rest, could do little more than throw himself into the wire as the guards came stampeding out of the duty hut, mad as hell. Seen as one of the ring leaders, he came in for some rather rude treatment, first, on the receiving end of Lieutenant Kelly's club and then, later, in Colonel McNally's office, since the colonel somehow felt that he must be partly responsible. Still, Jack was pleased to hear that the colonel, once he had calmed down, considered the break a clean one, and he could only agree with the colonel in marvelling at its ingenuity (an observation the colonel also conveyed to others, apparently), feeling proud of his best friend and knowing that Keefer at least would soon be back on ops, having the time of his life.

But that didn't mean Jack was going to sit there and take it.

So, a week later, as he and Girdlestone languished about in the compound, kicking sand around like real POWs, – this time their parole was suspended for nine days – the two happened to notice that the Irish Army medical orderly who had assisted them in the hospital looked a lot like Stanley Karniewski, their Polish colleague. The orderly had been stopping by daily to attend to the many scrapes and contusions caused by the wire, with enough syringes, iodine, and painkiller to anesthetize the entire camp. A plan then sprung to mind – not the new plan, that would come later – but another, simpler plan. With a quick change of clothes maybe they could get the Pole past the first gate. Stanley could then inject the sentry at the main gate and they would all make a dash for it.[38]

"And what in heaven's name do you think you're doing?" asked the orderly some ten minutes later, as Jack began stripping the man of his clothes in the area of the courtyard behind the bicycle shed. Girdlestone, Covington, and Shaw, the next three patients in line, had pinned him to the ground. "You'll never get away with this – I'll tell the colonel, I will."

"Fine, asshole, tell the colonel."

The internees were a little testy that week, no doubt about it.

"I will, I will, I swear." The Irishman was flailing about on the floor, putting up a spirited defense.

"Good, you stupid bastard, swear all you want."

Once the man's trousers were off, Covington started looking for PO Karniewski. "Stanley, where's Stanley?" Covie shouted. Stanley Karniewski appeared a moment later. Covington then ordered Stanley to take off his clothes. As Girdlestone later noted, discipline in the Polish Air Force must have been strong, for soon the Polish officer was heading out towards the main gate in the orderly's suit, a bundle of bandages in one hand and a dish of dirty water in the other.

Arriving at the main gate with the orderly's white cap pulled down over his eyes, Stanley made the mistake of offering the sentry one of his characteristic grins, displaying several of his twenty-two gold-filled cavities. "Your're not the orderly," said Corporal Smeaton. "You're Pilot Officer Karniewski. What are you doing in the orderly's clothes?"

And that was the end of that.

And then there was the Dalton incident later that month - which can't really be called a plan either (and certainly not the plan), as it was more a last-minute product of circumstance as well. Like Diaper, Brown, and Virtue, Dalton was a little miffed over Jack's

38 The only version of this particular attempt appears in Girdlestone's 1943 article and in subsequent interviews with the New Zealander. That's not to say that the event didn't happen, only that Keefer was gone by then and there was no mention of it in the archives. Brady and Covington didn't remember it either. Had Jack been the one to write this book, he would have lacked the second corroborative source preferred by any responsible reporter.

magazine article, which he thought made light of their circumstances. Also, it would have been nice had Calder seen fit to mention them by name, rather than just talking about himself and Keefer all the time. Jack was feeling a little guilty about that and after chatting with Dolly in the courtyard prior to the latter's departure on parole, he was delighted to meet Jim Masterton, the former Olympic swimmer who had survived Fleming's crash, on the way in. Apparently the guard had taken what he thought was Masterton's form out from the drawer and returned it to him as cancelled – only it was actually Dolly's form. This was how Webster had gotten out, of course, when he had received Wolfe's form by mistake. So Jack quickly pocketed the form, signed himself out, and then passed the form on to Dalton, who was loping down the road with Les Diaper for a beer at Osberstown House. And soon Dalton had a free pass to Belfast instead.

Of course the Irish kicked up a big fuss – even going so far as to draft a new form, tightening up the language yet again. Too bad, their mistake. Although, once more, the event would seem less fortuitous with Dalton dead within the year – in his case crashing a Lancaster III south of Munich near the Austrian/German border. Which raised the question: wouldn't they be better off just staying put?

A question which no one, to their credit, seemed prepared to pose.

No, the new plan was more subtle than that, more subtle than either the Karniewski or Dalton capers, and certainly more subtle than the CP press releases which Jack had "arranged" that week announcing Keefer and Fleming's escape from the Emerald Isle. Such fun to be a reporter back then – helping your buddies out at every opportunity and making your enemies, in this case, the censors, look like fools.

It would be a plan that had never been tried before, and likely would never be tried again. It would be a plan unique in the annals of military history, much like their internment.

It came into existance that week, Jack's plan, though its ethereal, understated quality makes it difficult in hindsight to ascribe it to a specific day. But there was something about the *Maclean's* article and everyone's reaction to it that struck Jack as odd. And that got him thinking.

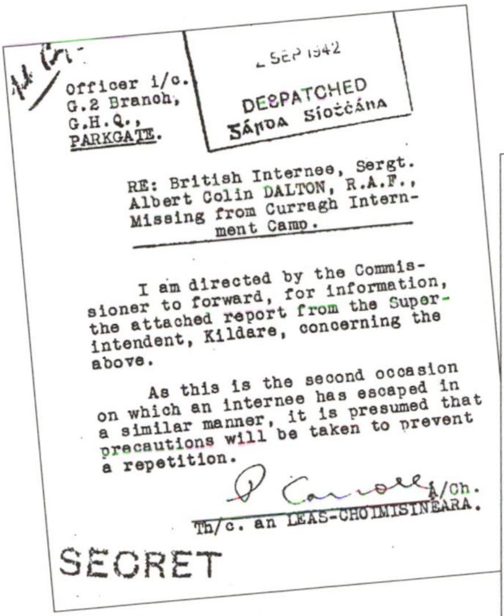

Officer i/c.
G.2 Branch,
G.H.Q.,
PARKGATE.

2 SEP 1942
DESPATCHED
Gárda Síochána

RE: British Internee, Sergt. Albert Colin DALTON, R.A.F., Missing from Curragh Internment Camp.

I am directed by the Commissioner to forward, for information, the attached report from the Superintendent, Kildare, concerning the above.

As this is the second occasion on which an internee has escaped in a similar manner, it is presumed that precautions will be taken to prevent a repetition.

A/Ch.
Th/c. an LEAS-CHOIMISINÉARA.

SECRET

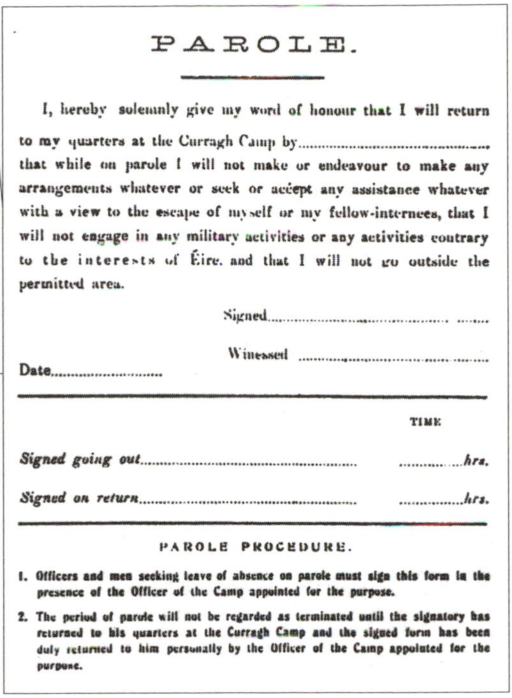

PAROLE.

I, hereby solemnly give my word of honour that I will return to my quarters at the Curragh Camp by.. that while on parole I will not make or endeavour to make any arrangements whatever or seek or accept any assistance whatever with a view to the escape of myself or my fellow-internees, that I will not engage in any military activities or any activities contrary to the interests of Éire, and that I will not go outside the permitted area.

Signed..

Witnessed ....

Date..........................

TIME

Signed going out.. ....*hrs.*

Signed on return.. ....*hrs.*

PAROLE PROCEDURE.

1. Officers and men seeking leave of absence on parole must sign this form in the presence of the Officer of the Camp appointed for the purpose.
2. The period of parole will not be regarded as terminated until the signatory has returned to his quarters at the Curragh Camp and the signed form has been duly returned to him personally by the Officer of the Camp appointed for the purpose.

Paragraph two of the Parole Procedure was changed on the new parole form, so as to continue the period of parole until the officer in charge had actually placed the cancelled form in the internee's hands, (compare the wording on page 168). This was meant to put an end to the internees running off before the guards could catch them and then claiming they could lawfully escape, something which irritated Colonel McNally to no end. Courtesy Irish Military Archives

It wasn't just what the Irish would do if he wrote something unfavourable next time. He'd thought of that before, of course, and there were plenty of English reporters slamming the Irish from across the Irish Sea, as it was. On the other hand, it wasn't enough to just make light of the place, judging from the fact that Colonel McNally had quite enjoyed the *Maclean's* piece, (especially the reference to the Irish kings near the end). No, Jack knew that he would have to come up with something more substantial than the self-righteous condemnations of the English press or the fun and games of "I Flew into Trouble" to get him kicked out of the place, to join Keefer back on ops as he so badly wanted. And looking at the bails of wire that had gone up in the past few weeks, no one was going to get out any other way.

No, it would have to be something more along the lines of what he sensed from the colonel after that, he decided, after the colonel had begun asking him questions, inquiring after the book at every opportunity, seeing the remarkable breadth of humour and insight of a man still in his twenties who had only been on Irish soil a month (at that time), yet whose views and ability to express himself immediately caught the colonel's attention. Sure, they were couched in stereotypes, as most unitiated views of the Irish were back then. The colonel knew that. But he immediately saw the potential, he admired the intellect. And now the former reporter was writing a book on Irish history, by Jesus, and he was serious!

As for Jack, he didn't *really* want to write a book about Irish history. Who would ever want to do that? Sitting on your bottom all day, your eyes bleary and your back killing you, drinking too much and ignoring your other responsibilities; missing all the action. But now that fate seemed to be propelling him in that direction, he'd have to learn even more about the place, he realized, so that no one – Colonel McNally, or Sir John Maffey, or the RAF – would ever dismiss him, or his views, again. He hated that, almost as much as he hated some picky, finger-up-his-bum editor sticking his stories in the back of the first section, with the socials.

He had to make himself a threat, he realized, and use the power of the press to his advantage. Otherwise he might never get out.

So, as the weeks passed, Jack continued his studies even more intently, with fresh purpose, throwing himself into the project, seeking to broaden his knowledge of Ireland at every opportunity – its religion, its politics, and of course its wonderful and illustrious history. That, in particular, he came to truly admire – careful, as always, to leave out the latest pieces of foolscap on his bed for the colonel's review. Soon he had completed the first eighteen centuries of his arduous journey, picking up far more detail than he would ever need for an escape plan – or a book, for that matter. From Cromwell, Drogheda, and Hell or Connaught to the English Penal Law of 1703 that had denied Catholics the right to vote. From Wolfe Tone and the United Irishman's Party – chanting *Erin Go Bragh* as he awoke each morning, not because he disapproved of how the English had treated the Catholic majority but because he was now over the hump with only 200 hundred years to go – to the Act of Union joining Ireland to the Empire. From Daniel O'Connell and the Irish Republican Brotherhood to Charles Stuart Parnell and Irish Home Rule. In fact, Jack had just started on Parnell when Colonel McNally paid his first personal visit, coming by the place where Jack now spent most of his afternoons reading and typing, in his shabby, lonely hut, often declining parole, grateful for the books that the colonel promised to bring him, shaping his mind. It was a great honour really, a personal visit from the commandant, almost as if he, the colonel, had taken Jack under his wing, like a father would a son. Who wouldn't want to usher an aspiring writer through the final stages of what any sane person would view as a tortuous, unfulfilling undertaking?

The plan was working.

Using his many contacts, including those acquaintances he had met at the Derby before the escape, Jack then turned his mind to Irish politics. As he knew, the squeaky wheel inevitably gets the grease and, while history was important, he would be a fool not to focus on current events if he really expected to be taken seriously, if he really expected to be unceremoniously kicked out of the country and sent back to war. Otherwise, they'd just think of him as some detached, half-crazed academic.

And, as luck would have it, there was an election that spring. It would pit de Valera's majority government against the dissidents in the fine gael, including James Dillon, the horse he'd be backing – though Jack was disappointed to learn that fall that Dillon had been kicked out of the fine gail for publicly urging Ireland's direct participation in the war. That worried him since he had genuinely hoped that there was a real issue there; that the pro-involvement side actually had some support. Maybe he should conduct more interviews, if not with Dillon, then at least with some other leading figures so he could present a more "balanced" picture. For a balanced picture, as any reporter will tell you, is always more difficult to dismiss than an unbalanced one.

This led to an introduction to Dan Breen, the following week, on Jack's next trip to Dublin. (As of October, trips to Dublin were permitted weekly.) Breen was a well-known republican hero who had since retired from politics and turned to business. He answered Jack's many questions as they stood outside the Dail, a large white building not unlike the White House, built in 1745 for the Earl of Kildare and later sold to the Royal Dublin Society prior to partition.

It housed both the Dail Eireann and Seanad, or Senate, roughly equivalent to the House of Commons and the Senate in Canada. They had a proportional representation system in Eire, Jack learned, one in which de Valera currently enjoyed a majority position, but if the Labour party, the Independents, and the Farmers made inroads, a minority government might result. When Jack suggested that Dillon's views might well be influential come May and June, Breen replied that almost everyone in Ireland, apart from Kildare, and parts of Monaghan, Dillon's home constituency, favoured neutrality, and that Dillon was spouting hot air out of his ass, or something to that effect.

That got Jack going. "How can you say that, sir? How can you?"

"I just said it, didn't I? Are you going to write it down or not?"

Jack put the pen back in his pocket, and closed his notebook. "No sir. I am not going to write that down. Now, I appreciate your past, being a great hero of the republican cause, and I understand all of that, truly I do sir, but surely the Irish would favour joining the war now that the tide has turned. The Germans are awfully busy in Europe, and the chances of Hitler attacking here are pretty remote in my opinion."

"In your opinion, lad? Well, isn't that nice. Can we rely on your opinion when the bombs start dropping?"

But then Jack realized that he had to restrain himself. To be neutral. Otherwise they'd never take him seriously. Otherwise he'd never get out. So he took the pen and pad back out of his pocket.

"Fine, Mr Breen, I can see your point," he replied, scribbling a few notes, omitting any references to body parts. "What are your thoughts, then? Where do you see the election going? Will de V win a majority this time?" (Good thing Keefer had left.)

"I don't give a hang for the election, Calder. I've told others that and I'll tell you that. I've retired from politics. Didn't Colonel McNally tell you that?" Breen smiled. He couldn't help but like the spunky, somewhat arrogant reporter. He'd done the interview as a special favour for a friend in the government; humour the poor soul. "Still, if you're serious about this you might want to focus on the positive aspects of your situation. Parole, for one."

"Sir?"

"You get parole don't you? What are you complaining about?"

"I'm not complaining, sir. I'm investigating."

"Fine. Keep investigating, but in the meantime tell the world about what a good sport Ireland has been."

Of course Jack thought about that exchange on his way back to the Curragh on the bus that night, passing through Rathcool, Newbridge, and Naas, missing his two good friends more than ever. He hated it when people thought that it wasn't tough being stuck there, as if they weren't real prisoners. He could, he supposed, tell everyone what good sports the Irish were, since he didn't really dislike the place at all, although his final answer to Breen, a snarky rejoinder to the effect that how he presented the article would depend on how much longer they kept him there, wouldn't have left that impression. Still, they both agreed in the end that slamming or denigrating the Irish as the English seemed inclined to do accomplished little. Breen went on to suggest that it might be fair, or "balanced," to mention the contributions and accomplishments of the great Irish generals who seemed to be overlooked by everyone, names like Captain Fogarty Fegen, Cdr. Eugene Esmonde, and Captain H.M. Irvine-Andrews, names Jack diligently wrote down just in case. That ought to make the government good and nervous, he thought later on the bus, a short piece on Irish heroes of World War II.

So, in short, while Keefer and Fleming, his good friends, were off having the time of their lives (he assumed), Jack's new plan was taking shape, vague at first but clearer and clearer as the weeks passed. While he was undoubtedly disappointed that he hadn't been over the top the previous summer when Keefer's curtain rod gave way, he was hardly throwing in the towel.

When Keefer and Fleming arrived back in Canada, it wasn't as either would have preferred. Hailed as conquering heroes on the one hand – thanks to Jack, who seemed awfully active for someone supposedly retired from his profession – and the butt of more than a few jokes on the other – frightful business, being an Irish internee, care to talk about it? – they had been sent home for a "well deserved rest." So they were almost as disappointed as Jack. Expecting as escapees to write their own ticket, they had prepared their own separate wish lists on their arrival in London. Keefer was hoping for a switch to daylight flying – a stint in photo reconnaissance, maybe, flying either Spitfires or Mosquitoes – while Fleming was hoping for his own four-engine Liberator. But neither expected what actually happened.

"Yes, well, normally escapees can write their own ticket, but you lads ... I'm not quite sure what to make of you."

That was the line they were given by the Squadron leader who interviewed them in London. RAF Ferry Command would be setting up shop at the newly completed Dorval Airport in Montreal, and they needed instructors. When Keefer suggested that his 333 hours fell well short of qualifying him for such an honour, the reply was clear:

"Keefer, if I post you to Dorval as an instructor, you will go to Dorval as an instructor."

So that was that. As compensation they were granted a whole week's leave in London, where they visited Park Lane and Minsky's, bumping into old friends. They also had dinner with Uncle Wilks at Corries, and with Dode and Blink, the two girls who had accompanied Keefer and his navigator on their cross-country adventure to Scotland. They both liked Grant but missed Jack, particularly Dode. Keefer passed on Jack's regards, of course, telling her how much Jack missed her; how he had not really met anyone in Ireland who could match her beauty and charm. Sure, Jack had met a few women there, but none nearly as intelligent as her. It hadn't really been his fault that the Hudson had flipped over in the mud outside Hadrian's Wall, referring to their unfortunate mishap on the way to Scotland that day, its pencil-thin tires spinning helplessly towards the sky. Keefer was driving, after all. Jack's strength was in navigating, they all knew that. And while he wasn't about to make any promises, since he wasn't sure if he was going to survive his internment, if he ever did get out of Ireland, past the wire and the guards, she'd be the first one to know.

Before long both Keefer and Fleming had settled into their new lives in Montreal, in separate apartments on Sherbrooke Street. It was an important job, Ferry Command, so they should have been proud even if they weren't. The Dorval depot was used to try out new engines and train new pilots. While American and Canadian factories had been busy manufacturing planes for overseas operations, many of these planes had been sunk or destroyed in naval transit by German U-boats. So trans-Atlantic flights were the only real alter-

P.O. Bob Keefer.

P.O. Robert Keefer Native of Ottawa, Escapes From Eire

Relative of Rockcliffe Resident Was Interned When Plane Crashed Last Year.

The flying feet that made him an outstanding football player on the McGill University team have carried Pilot Officer Robert (Bob) Keefer, a native of Ottawa, out of the reach of internment officials in Eire, according to word received here. Pilot Officer Keefer is a second cousin to Allan Keefer of Rockcliffe and a son of Mrs. Keefer, of Montreal and the late Edward Keefer.

How Pilot Officer Keefer effected his escape from an internment camp is not known here, but he is safe, presumably somewhere in England. His brother Ralph is also an officer with the Royal Canadian Air Force in England, and played on the same McGill team. The brothers enlisted together at the outbreak of war and have seen considerable action. It is believed that Pilot Officer Keefer was taken prisoner and interned in Eire when his plane crashed in that country last year.

The Keefer boys were born here and moved to Montreal with their parents at an early age. Both, however, have visited Mr. and Mrs. Allan Keefer at Rockcliffe on occasions, and have a number of friends in Ottawa.

Keefer Plunges to Liberty

Former McGill Football Star Escapes Internment in Eire

Calgary Flier Also Gets Away With Local Airman From Camp

TORONTO, Sept. 1—(C. P.)—P/O. Bob Keefer of the R.C.A.F., one-time McGill University football player, has escaped from internment in Eire where he had been confined since his bomber crashed last year, the Canadian Press learned today.

Word of Keefer's escape came soon after a Calgary dispatch reported that Flt.-Lt. Grant Fleming of Calgary, interned in neutral Eire since his bomber crashed last winter, also made his get away. Fleming's escape was reported in a cable to his cousin, Jack Fraser.

F/O. R. G. Keefer

Fleming won the D.F.C. for piloting a crippled flying boat over the Atlantic last summer.

Keefer was pilot of the plane in which Jack Calder of Goderich, former Canadian Press editor in Toronto, was navigator.

This break-away followed an unsuccessful attempt early in the summer of Canadian and English airmen to flee from the Curragh interment camp. Whether Fleming was a participant then was not known.

Names of the participants never were made public but stories reaching Rathdrum, over the hills from the Curragh in County Wicklow, told of a rough-and-tumble attempted mass escape which came to grief because of one thing—"Ersatz" bolts in the ladder joints. When the ladders collapsed the guards rushed in.

Among the Canadians known to be at the camp with Fleming are F/O. Jack Calder, Goderich, and Sgt. Pilot Paul Webster, Vancouver. Sub-Lt. Bruce Girdlestone, a New Zealander in the Fleet Air Arm is also interned.

THURSDAY, OCTOBER 1, 1942

Reunited After Fleeing Internment

Flight Lieut. J. Grant Fleming, D.F.C., of Calgary (left) shown with Flying Officer R. G. Keefer, of Montreal, reunited in Canada after escaping from an Eire internment camp. Fleming and his gunner swam four miles to shore when his flying boat crashed off the coast of Ireland with a loss of nine men. Keefer and his crew were forced to bail out of a Wellington bomber over Eire.

CP wire stories arranged by Jack Calder, following Keefer and Flemming's successful escape. Clockwise from left: *Ottawa Citizen*, *Montreal Star*, *Calgary Herald* and *Globe and Mail.* The *Citizen* had a number of details wrong as usual, sources of information being what they were during the war. First, Keefer's given name was actually Ralph, only he had the good sense to pick up a nickname along the way. Second, while his brother Eddy played football, they never played on the same team, nor was Eddy a flyer, choosing the navy instead. And third, Keefer's parents, or mother since his father had long since died, never moved to Montreal, though he did. Courtesy CP

Two Canadian War Fliers Escape From Irish Camp

TORONTO, Sept. 1.—(CP)—Pilot Officer Bob Keefer of the R.C.A.F., one-time McGill University football player, has escaped from internment in Eire where he had been confined since his bomber crashed last year, the Canadian Press learned today.

Word of Keefer's escape came soon after a Calgary dispatch reported that Flight Lieut. Grand Fleming of Calgary, interned in neutral Eire since his bomber crashed last winter, also had made his get-away. Fleming's escape was reported in a cable to his cousin, Jack Fraser. Fleming won the D.F.C. for piloting a crippled flying boat over the Atlantic last summer.

Keefer was pilot of the plane in which Jack Calder of Goderich, former Canadian Press editor in Toronto, was navigator.

These escapes followed an unsuccessful attempt early in the summer of Canadian and English airmen to flee from the Currah internment camp. Whether Fleming was a participant then was not known.

Names of the participants never were made public but stories reaching Rathdrum, over the hills from the Curragh in County Wicklow, told of a rough-and-tumble attempted mass escape which came to grief because of one thing — "ersatz" bolts in the ladder joints. When the ladders collapsed the guards rushed in.

native. But that was still new, especially in winter when icing was a problem. To fly such missions, one needed skilled pilots. So it was important not only that the aircraft be up to snuff but that the pilots be trained properly. Otherwise, neither would ever get into combat. Still, Keefer found it boring.

Not that he didn't find ways to alleviate the boredom, mind you, developing some of the same almost sadistic pleasures of other instructors before him, seeing young kids screw up just short of death and disaster. "Recovering from unusual positions" is what he called it, banking the aircraft, (in this case, a Hudson) sixty degrees and then leaving his seat and heading to the back, leaving the kid alone at the controls. Or getting the kid all set up with his full instrument panel and then reaching down and quietly shutting off his fuel. That was fun. There'd be an instant yaw as one engine would cut – he'd never do two at the same time, he wasn't stupid – before the kid would start pounding away on the rudder pedals like a pump organ. Then the poor sucker would fumble with the rudder trim, checking the gear and flaps, and all but peeing his pants while Keefer rolled his eyes, sighed impatiently, and did all in his power to make the sprog feel like a complete dolt. Then, finally, as Keefer quietly feathered the engine and took them out of the tail spin, the kid would catch on. They hated that. And they hated him too. That was the plan of course, in the hope that he would eventually be sent back to the war, remembering how his father had died, in Virginia, in exactly the same position. These kids were dangerous. He'd be safer over in Europe!

Still, life wasn't all bad. Maybe he should make the best of the situation, he thought, just as he and Jack had in Ireland, having learned from their English colleagues: that when all else failed, there was no point worrying.

Except that in late December the two were told they'd be kept in Ferry Command for a further six months. They had assumed, erroneously, that they had done their penance, though no one ever expressed it that way. Fleming would be sent to Elizabeth, New Jersey, to become the commanding officer of the flyingboat base there. Keefer, meanwhile, would be on his way to North Bay, Ontario, which in winter would make Ireland seem like Hawaii.

As for the rest, Girdlestone, Covington, and the others, having failed in their attempts to go over the wire in February, and through

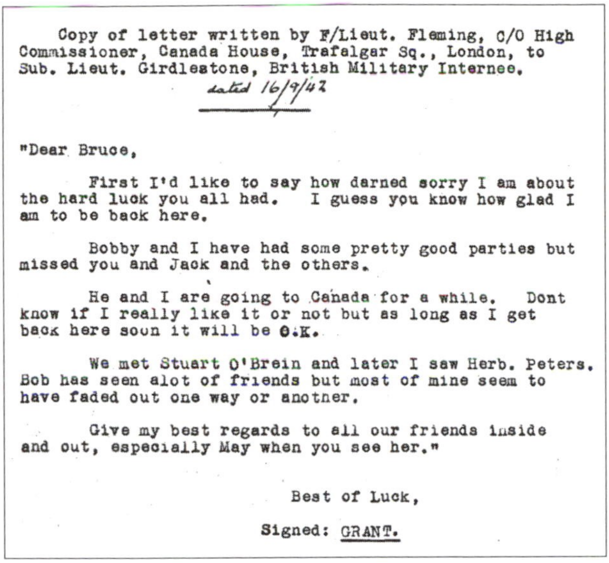

Copy of letter written by F/Lieut. Fleming, C/O High Commissioner, Canada House, Trafalgar Sq., London, to Sub. Lieut. Girdlestone, British Military Internee.

dated 16/9/42

"Dear Bruce,

First I'd like to say how darned sorry I am about the hard luck you all had. I guess you know how glad I am to be back here.

Bobby and I have had some pretty good parties but missed you and Jack and the others.

He and I are going to Canada for a while. Dont know if I really like it or not but as long as I get back here soon it will be O.K.

We met Stuart O'Brein and later I saw Herb. Peters. Bob has seen alot of friends but most of mine seem to have faded out one way or another.

Give my best regards to all our friends inside and out, especially May when you see her."

Best of Luck,

Signed: GRANT.

Above, letter from Fleming to Girdlestone, 16 September 1942, intercepted by the camp censors and copy-typed. It is unclear whether the Irish ever suspected May. They were aware of "Tipperary" who was probably Michael Burke, a well-known IRA member. But May is never mentioned in any of the reports. Courtesy Irish Military Archives

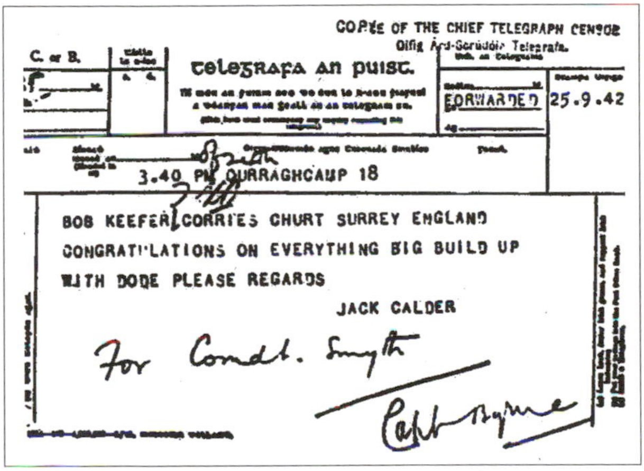

COPIES OF THE CHIEF TELEGRAPH CENSOR
Oifig Ard-Scrúdóir Telegrafa.

Telegrafa an Puist.

FORWARDED 25.9.42

3.40 PM CURRAGHCAMP 18

BOB KEEFER CORRIES CHURT SURREY ENGLAND
CONGRATULATIONS ON EVERYTHING BIG BUILD UP
WITH DODE PLEASE REGARDS
JACK CALDER

For Comdt. Smyth

Capt Byrne

In fact, Keefer, who was inclined to build up his friend at a moment's notice in any event, never received this telegram. Whether that was the work of the chief telegraph censor remains uncertain. Courtesy Irish Military Archives

it in August, they decided in December, over Christmas dinner, to go under it. They would build a tunnel. Not Jack, he was too busy on his book, and not that night, not as they all enjoyed a splendid holiday feast prepared by their three Polish colleagues, Pilot Officer Stanley Karniewski, Flight Lieutenant Kazimierz Baranowksi, and Pilot Officer Jan Zimek, a new internee who had arrived the previous month. (Stanley had done the menu, the Baron had done the cooking, and Zimek had done the complaining. "Why me?" Zimek kept asking, having become the first Allied flier to be interned since Girdlestone a year earlier, when all the others were being shown to

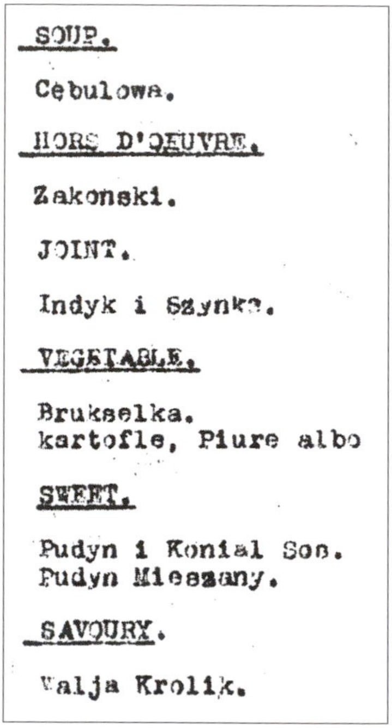

SOUP.

Cebulowa.

HORS D'OEUVRE.

Zakonski.

JOINT.

Indyk i Szynka.

VEGETABLE.

Brukselka.
kartofle, Piure albo

SWEET.

Pudyn i Konial Sos.
Pudyn Mieszany.

SAVOURY.

Valja Krolik.

Don't ask me what any of it means, he'd say. Courtesy Bruce Girdlestone

the border.[39]) But within the next month or so, when all the shovels and bags had been readied and a proper plan put in place. They certainly had nothing to lose.

The first step in building the new tunnel, strangely enough, was to acquire a dog. While Jack wasn't directly involved in the tunnel – he was so wrapped up in his studies that he didn't even know about it – he was, in fact, responsible for the dog. As Colonel McNally was a keen breeder of pointers, and the two were becoming such fast friends, the colonel offered Jack one from the next litter. While Jack could have cared less about a dog, he did mention the offer to Girdlestone, whose fertile imagination then went to work. A puppy would require a kennel, which could provide a suitable front for a tunnel entrance.

Sure enough, after the puppy was born, it was delivered to Camp B by one of the colonel's NCOs. Unfortunately, it became a terrible nuisance in the yard, barking at the guards, and finally urinating on them, as if it had actually been trained to do that. Girdlestone then went and spoke to the colonel. He managed to convince the colonel that the best way to solve these problems was to build the dog a proper kennel. And the most logical place for the kennel would be under one of the huts, given the two or three feet gap between the floor and the earth beneath it. This way the dog would be out of sight. In the spirit of benevolent neutrality, Colonel McNally kindly agreed.

39 While estimates of the number of Americans who were quickly and quietly escorted north during that period of the war vary from two to several dozen, one of the more noteworthy incidents definitely occurred that month. It involved the force-landing of a four-engine Flying Fortress on its return from North Africa, following a secret reconnaissance trip in advance of the Allied invasion later that year. All sixteen American flyers involved were released, including General Jacob L. Devers, chief of the US Army Ground Forces. The release was authorized by F.H. Boland, secretary of the Irish Department of External Affairs, on de Valera's order.

So, soon they were busy at work building the kennel. After discussing the matter with Commandants Mackay and Guiney, the colonel allowed them to use some of the timber left over from the earlier construction for the walls. Sergeant Duncan Fowler, from Victoria, British Columbia, then jumped into action, putting his Diston Cross-cut saw to good use. And before long the kennel's walls were up.

Using the kennel as camouflage, they then began the next home-construction project – the tunnel. First, they dug a hole, covering it up at night with some turf and an old Guinness crate from the officers' bar. As a rough guide, they'd have to go nearly one hundred feet to be beyond the main gate, a daunting task. Still, with the gardening forks they had smuggled into camp, it was possible. And it had to be better, they agreed, than spending every day on parole, or every night in the officers' bar, or vica-versa.

The kennel had been designed to have two entrances: one from the front, at ground level, labelled "Kennel," and one from above, through a trap door that led directly into Girdlestone's hut. However, one night when Fowler was working overtime finishing the kennel (like many of them, Fowler was ignorant of the real purpose of the kennel), they discovered that the opening between Girdlestone's floor joists was too narrow to allow anyone to pass through. This meant that they would have to move the kennel and start the tunnel all over. Fortunately this wasn't too much of a burden, as they had just started. The problem was solved the following day by moving the kennel to the area beside the bicycle shed.

By early February digging had resumed, in the area immediately between the wash house and the main hut where Remy slept. By then Remy had been brought in on the plan, since the entrance was now through his bedroom. The distance this time was a little more daunting – nearly two hundred feet to the main gates, and nearly one hundred feet to the first fence to the east. The Germans lived over there, of course, to the east. And they certainly didn't want to tunnel into *their* camp. Not one at a time anyway.

The next step was to set up an alarm system. They had brought in two others for the digging, Brady and Harkell, an English sergeant, and they needed a way to warn the group in the event of any surprise inspections. So they rigged up a stone pendulum on the end of a chord and ran it into Remy's room, since he had the best view

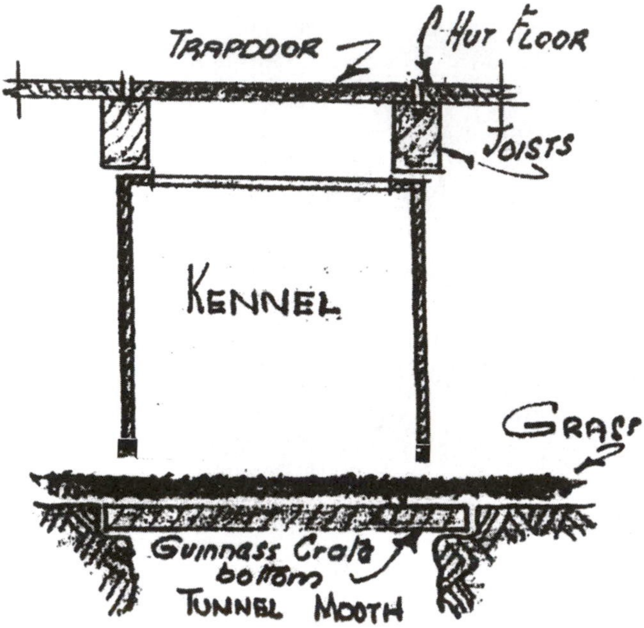

Girdlestone's drawing of the tunnel entrance from his 1943 report on the camp. The width of the joists below Girdlestone's hut was only twelve inches, so it wasn't surprising that they had to move the tunnel to the bicycle shed. What was surprising was that the guards never figured out why.

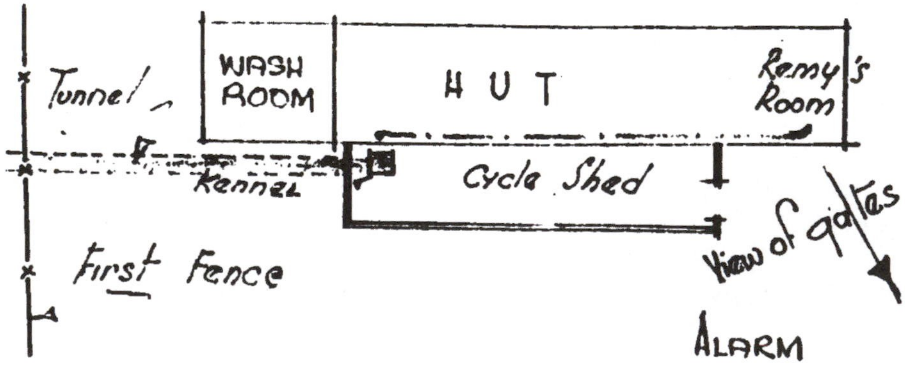

Again from Girdlestone's 1943 paper on the camp. Remy's room was located in the hut immediately behind the mess. Courtesy Bruce Girdlestone

From left, Girdlestone, (the best man) Rev. Flint, the groom, and the bride, Nora's brother and mother. Courtesy Bruce Girdlestone

Covington, Girdlestone, Rev. Flint.

of the main gate. A few pulls on the chord would send the pendulum into action, striking the turf with a thud, which would warn the diggers that someone was about to enter the shed.

So far so good, even if progress was painfully slow – about a foot a day – meaning that if all went well they'd reach the main road in roughly two hundred days. They could only dig before lunch or after tea, unfortunately, and had to carry the soil out in twelve inch cloth bags, spreading it slowly and carefully about the yard. And there were other, less conventional, problems. First, they were interrupted by Covington's wedding. It was tough to make any headway digging a tunnel, Girdlestone noted many years later, with prisoners taking off every few days to get married. That was an exaggeration of course – there were only five weddings that year.

The Covington wedding was planned for a Saturday. However on the Tuesday, after a spat with Nora's parents over the ceremony that arose from the fact that it was to be the camp's first mixed marriage, Covington had to send four NCOs over to her house on Friday night and literally kidnap her. She was then taken to the Brownstown estate where Ward lived – though Covington to this day maintains that the operation was more in the nature of a rescue than an abduction. Anyway, the two were married the following morning in Newbridge by Reverend Flint in a Protestant church ceremony attended by her mother and brother, though not her father. So that held things up.

And then Zimek, the Polish-American pilot who had landed earlier in the year, got into a nasty fight with two German flyers. He took offence at a friendly conversation between Karniewski and a couple of Germans at a dance hall in Newbridge one night and, after a dispute over a girl with another German a few nights later, two Germans decided to get even, jumping him on the way home. They nearly killed him.[40] Parole was immediately suspended for everyone and only reinstated on alternate days for the first month after that. The Irish were concerned that the dispute might develop into a war between the two sides, something they had so far avoided, and were anxious to avoid in the future at all costs. As for the tunnel, once parole was suspended, with so many people around

40 The Zimek incident is described in some length in T. Ryle Dwyer's book, *Guests of the State* (Ireland: Brandon Books, 1994).

they had to stop digging. Mostly because of Ward, who would be the first to turn them in if it meant a promotion.

And then, last but not least, a reporter arrived in camp. Not Jack, another reporter this time, a real reporter. A reporter who wanted to interview some internees, including Jack, Wolfe, and, of course, Covington. That took some time and set them back, too.

The arrival at the Curragh of Sam Boal of the *New Yorker*, one of the world's great magazines, in the last week of March 1943 complicated things further. Not that it meant that Jack's plan would no longer work, or that he would have to come up with a new one. Certainly not. Albeit a bit fuzzy around the edges, the plan was still something to be reckoned with, still solid in the middle. As a multilayered, and creative solution, it would not be defeated by one intervening act of fate. Indeed, Boal's visit might work to Jack's benefit, having another writer around, though he hadn't exactly worked out how. Only that Boal's appearance complicated things.

The reason, of course, was that Jack had just finalized arrangements with Canadian Press for his next dispatch. It was to be a five-part exposé of the camp, his editor had decided, a real barn burner. Jack would explain to the Canadian people all about Irish neutrality, since there seemed to be a bit of confusion about that. No one could quite figure out why the Irish had chosen to remain apart, especially now that the tide had turned. He would also touch upon Irish history and politics, the upcoming Irish election, and finally something of a general nature about Irish customs and temperament, going beyond the generalities, the stereotypes to expose the incongruities, the paradoxes – the Irish temperament. The articles, which would begin in late May or early June, were expected to be hard-hitting yet sensitive and insightful. Difficult to dismiss. Canadian Press was excited. It would be the first serious examination of the Irish character and temperament by a non-Irish writer since George Bernard Shaw – who may have been Irish but certainly got disowned a lot over there. Jack had already sent the first two instalments to Toronto prior to Boal's arrival, receiving in return an advance, care of his account with the Royal Bank in Goderich, Ontario. They'd be furious, finally encountering a foreign reporter whom they could not merely dismiss. He could see it now: the first North American journalist to report on Irish neutrality in an objective way from where it mattered most, peering through the wire. If he wasn't out on parole, that is.

Only now Jack might get scooped. And there is nothing worse for a reporter, even a dispassionate and independent one, than getting scooped. How could he look his friends at CP in the eye again?

A REPORTER AT LARGE

THE GREEN AND LEGAL CURRAGH

The variety of nationalities of the Allied fliers may have an emotional importance, but it has no legal importance whatever. All of these fliers were acting as agents of the British crown when

Sam Boal's article in the *New Yorker*, published less than a month later on 15 April 1942, was all that Jack feared. A funny, hard-hitting piece that caught the many ironies and paradoxes of their situation, it made a lot of the points Jack wanted to. Indeed, it blew the cover right off of their internment! What a ridiculous situation! Why don't the Irish just release the poor fliers? At least that was the inference that they, the internees, drew from it.[41] Why rely on all the legal niceties when no one in their right mind actually believed that the Germans would ever attack Ireland this late in the war?

How was Jack going to get himself kicked out of the country now? Now that Boal had stolen his thunder?

Though, in truth, Jack's plan was never that simple. It was never merely to get himself kicked out of the country by making such a nuisance of himself that he would become a threat to the internal security of the Republic, which was still very much a nascent Republic, in a very difficult period of its history. Hell, they'd just lock him up down the road with the IRA. That would have been too easy, and more important, the Irish would have figured it out almost immediately. Those guys are smart, the Irish, and, make no mistake about it, Jack wasn't taking on just Colonel McNally, or the adju-

41 That, and the fact that it made too light of their situation, at least according to Brady, Covington, and Wolfe years later. Girdlestone, on the other hand, was more inclined to see the humour of their situation. As for Jack, my guess is that he would have agreed with Girdlestone.

tant general in Dublin, or Sir John Maffey for that matter. He was taking on the entire country from de Valera to Dan Breen, from Daniel O'Connor to Michael Collins, from Wolfe Tone to the great writers and poets and playwrights of the age like Yeats, Beckett, Joyce, and Shaw,[42] men who, in his simple estimation, had made the Irish what they were – that odd mix of sentimentality and excitability Girdlestone noted years later. After de Valera, that playboy of the Western World, had long since retired from the world stage. After Synge and the Dublin riots, when excitability had won out in the end. And finally after Oscar Wilde, who deserved a far better fate, had so aptly noted that the only way to deal with temptation was to yield to it. Like a writer getting up in the middle of the night when he can't sleep, or an officer staying up for one more glass of port at the officers' bar after the Royal Anthem had sounded.

How could he yield to temptation now with all the world at war, and the bloody Irish sitting on their fat asses, doing nothing? How could they do that to him? How could the bastards not understand?

No, Jack had something else in mind. He just wasn't sure if it would work.

When the first instalment of Jack Calder's five-part series appeared in Canadian newspapers on 9 June 1943, the former navigator, aspiring author, and now bona fide feature correspondent was nowhere to be seen. He was holed up in his room, refusing all visitors, including Colonel McNally. Moody and morose, he had turned inward. Already drinking to excess, he was now consuming a full bottle of brandy a day, sometimes two, occasionally three, literally splashing it down his throat. He had even refused parole, which hardly pleased Ann Mitchell, the woman he had expected to marry, having sent her long, flowering letters from camp (via the camp censors) letters fast and furious, thick and heavy with passion and commitment. More often than not he refused to talk – to anyone. And then, just as suddenly, he would turn with a vengeance on whoever was present, arguing vehemently until he was red in the face, ready to explode.

It was a difficult time, and Jack's strange behaviour was begin-

42 Or the fifth and last Irishman of the twentieth century to receive the Nobel prize for literature, Seamus Heaney …

" … between my finger and thumb
the squid pen rests, snug as a gun."

The first of the five articles, left, appeared in the *Toronto Star*, 9 June 1943. Other versions with different headlines appeared in most Canadian dailies, including the *Ottawa Citizen*, the *Calgary Herald*, the *Vancouver Province*, and the *Montreal Star*, each of which retain microfilmed copies on file from the war years.

INTERNED CANADIAN FLIERS LIVE LIFE OF RILEY IN EIRE

Frequent Parole Enables Them to Discuss Neutrality in Private Homes

JACK CALDER

C. S. BRADY

Jack Calder, former Canadian newspaperman, has been in an Eire interment camp for over a year. Frequent parole has given Calder a chance to observe Irish life in and around Dublin with a keen eye. In this, the first of a series of articles, he describes life in the camp and explodes some rumors about it.

By JACK CALDER

British Internment Camp, the Curragh, Eire, June 9—(CP)—**Thirty-four perplexed young men with hundreds of thousands of dollars' worth of flying experience, are almost living the life of Riley in the British internment camp in Eire—almost, but not quite.**

They are Englishmen, Scotsmen, Canadians, Poles, a Welshman, a New Zealander, a Fighting Frenchman, an American, and even a wireless operator from Northern Ireland.

The Canadians are Flight Lieut. L. J. Ward of Vancouver, the senior officer and a member of the R.A.F.; Flying Officer C. S. Brady, Toronto; Flight Sergt. F. W. Tisdall, Moncton, N.B.; Sergt. Roswell Tees, Thorold, Ont.; Sergt. Duncan Fowler, Port Alberni, B.C., and myself, all of the R.C.A.F. The American is Flying Officer Wolfe of Lincoln, Neb., a member of the now-defunct Third Eagle Squadron. A good few of us have Irish blood and Brady's father was born in Dublin.

We travel about the countryside on parole within set boundaries and freely discuss Eire's neutrality in public houses and private homes. Once a week we are allowed to leave our camp on the Curragh for a one-day visit to Dublin, 30 miles away.

Live Near Germans

Any attempt at breaking parole would be stepped on immediately by the British government, though once a man has re-entered the camp and got back his parole certificate he may try to escape.

Most of the fantastic stories which have sprung up about the camp have to do with our proximity to the German internment camp. A fence and a barbed wire barrier separate us. A mile away is the camp for Irish political prisoners. We never fraternize with the Germans, although they are accorded the same privileges as ourselves.

The legend that we sit "side by side" with the Nazis in church must have been derived from the fact that they appeared twice in the same church as the Protestant British internees last year. When some of the English lads helped the Poles with their shopping in Dublin, the story got to England that we had visited Dublin with the Germans.

All Liked Tommy

British troops garrisoned the Curragh until 1922. Tradespeople remember the Tommy as a "great spender" and the British officer as a hard rider to hound who "always paid up." There is probably a larger proportion of pro-British people about the Curragh than anywhere else in Eire. These facts make the internees welcome in scores of homes.

It is a relief from camp life to be able to spend a day working in the fields or felling trees, listening to the philosophy of the Irish laborers. Farmers like to have Sergt. Fowler around for a day because, with his experience of the British Columbia woods and Ontario farms, he sets a pace that their workmen must emulate to maintain their dignity.

One of the world's best-known race courses is only a mile from us. Outside our working time, we find diversion in going to the races, riding, swimming, fishing, football and walking or cycling through the green countryside. Three of the internees play rugby for neighboring Newbridge, three of us have played soccer for the same town, and our own soccer team plays Irish Army elevens regularly.

So friendly have we become with the horsemen that racegoers frequently approach us for tips. It would be a luxurious way of living through the war, for we remain on full pay. Yet the main topics in the messes are war and flying. At breakfast someone always relates a dream about escaping, or flying, or home.

Unfortunately this is no place to discuss escaping or escape attempts. My sorriest failure came when my great friends, Flight Lieut. Grant Fleming, D.F.C., of Calgary, and Flying Officer Bob Keefer of Montreal, got away last August, subsequently returning to Canada.

Since then miles of barbed wire has been added to the several fences. The other main obstacles remain the same—raised huts which prevent tunnelling, a small camp area and a force of intelligent guards handpicked from the Irish army. They know a thing or two about escaping as a result of their experiences in the "troubles."

There has been one recent consolation. We used to have to listen to "Deutschland Uber Alles" and shouts of "Heil Hitler" from the German camp on the nights of Axis victories.

When the start of Montgomery's march in North Africa was announced, we smuggled fireworks into our quarters. The young Nazis have never had a chance to reply to the hullabaloo we raised that night.

Irish Feeling Relieved Axis Being Hammered

There is joy in Eire at the hammering Britons and Americans are giving the Axis at last. She knows her future depends on Allied victory. Irishmen fight for the cause in every army but their own. Yet her ports are denied the United Nations in their battle against the U-boats. Air bases in Eire are not available to the Allies. And Canadian, British and American airmen, if forced to land within her borders, are interned for the duration.

In this article, second of a series, Flying Officer Jack Calder reviews aspects and paradoxes of Eire's neutrality.

By Jack Calder
(Copyright, 1943, by The Canadian Press.)

BRITISH CAMP, THE CURRAGH, Eire, June 10.—(C.P.)—Irishmen have begun to talk more freely about the war since the tide has turned in favor of the Allies.

Visiting in the Curragh-Dublin area on parole from Eire's British internment camp, I have found great relief at the fact that the Axis is being hammered at last. There is still a widespread inclination to look on the war as something between "England and Germany" and "none of our business," but the veiled fears and sympathies for the British cause have been uncovered today.

Neutral Eire has known right along that most of her aspirations for a new national life would be defeated in the event of an Axis victory. Despite the stern censorship, the people have realized the source of many of the things they need for living through the war.

England Can Give It

Premier de Valera has said that the British government has behaved "not unworthily" to Eire since the beginning of hostilities. Nevertheless, he has reiterated the policy of neutrality and cited the evident danger of an internal upheaval in the event of belligerency.

"Oh, England can take it and give it back, too," the Irish acknowledge freely now.

And beneath that tribute lies another main point in Eire's outlook. The people, not unnaturally, fear bombing.

The country's larger communities—Dublin, Cork, Limerick, Galway—are ports and would be immediate military targets in the event of entry into the war. A stick of German bombs across a residential district of Dublin two years ago caused sufficient death and destruction to leave its mark in the minds of the community. Worldwide emphasis on the ordeal of the English population served, in a way, to strengthen Eire's determination to avoid the horrors of modern war.

We internees, eager listeners to the possibility of any chance to get back to our duty of flying against the Axis, used to hear: "When America comes into the war, Eire will have to come in too. Then you'll be set free."

What They Thought

Japan attacked and the United States went to war with the Axis. For one brief period many persons with whom I talked thought it would be only a matter of time before they would be forced into war beside their old friends and close relatives in America.

Then Eire saw that America's involvement didn't involve her directly and, having heard from Mr. de Valera that her position was "that of a friendly neutral," breathed the pleasant air of noncombat more freely. United States troops arrived in Northern Ireland in January, 1942, and Mr. de Valera protested that he hadn't been consulted.

Still Eire came no closer to war. Today, with the danger of a German invasion of these islands apparently farther away than ever, war for Eire is farther away than ever, too.

Meanwhile the steady migration of Irish to join the British forces or take civilian war jobs in Great Britain has continued. The country takes pride in the achievements of Irish generals and admirals, in the award of the Victoria Cross to men like Capt. Fogarty Fegen, Cmdr. Eugene Esmonde and Capt. H. M. Irvine-Andrews.

Queer Machinations

Two stories, which I have heard again and again, are worth repeating to the people at home, as illustrating the queer machinations of Irish neutrality. I have heard the first one attributed to several Irish heroes, usually the late Paddy Finucane, and the second to half a dozen well-known Englishmen who have visited Eire since the outbreak of war.

Someone is supposed to have asked an Irish warrior why he should want to fight for John Bull.

"Ach, I don't give a hang for John Bull," the hero is supposed to have replied. "But I do like a bluddy good fight."

The story probably springs from the material in George Bernard Shaw's play "John Bull's Other Island." I find Shaw variously owned and disowned by the Irish according to his latest utterance.

In the other tale, an English intellectual was dining with a group of Irish savants in Dublin. He was asked what he thought of "partition."

"Partition?" the Englishman asked. "What partition? Never heard of it."

Someone, to whom the question of partition supersedes all others in the world, explained that Ireland has been cruelly divided into two parts, much against the will of all concerned, and that the division is all that keeps Eire from being the complete friend of Great Britain.

Irish Don't Like It.

It has become the custom for London newspapers to send men to Dublin for a few good meals and bring them back, somehow or other, to write their experiences of Eire's reaction to the war. The Irish usually don't like these reports.

"Even you aren't qualified, after all the months you've been here, to tell anyone what Ireland is like," they warn me. "Why, you haven't been to Ireland yet. You've just been to Dublin and the Curragh."

Then the speaker, depending on where his home is, tells me that I should go to Galway or Kerry or Cork or Limerick to see the true Ireland. Dan Breen—now a member of the Dail and a business promoter, but once the most hunted and most bullet-riddled of the fighters for Irish freedom—told me I should visit Tipperary to get the true Irish views.

My opinions must be accepted with the understanding that I shall probably leave Ireland without ever having seen it or found out what it was like. That will break my mother's heart, for she made me promise when I came overseas that I wouldn't go home without having seen the land of her grandparents.

Stirred Feeling Abroad.

But I do know that when James M. Dillon, a member of the Dail for the border county of Monaghan, urged direct participation in the war last year, he didn't get wild and enthusiastic support anywhere. The main result of his words was to stir anti-Irish feeling abroad, where he was widely quoted.

"Ireland's fate is bound up with that of Britain and the U.S.A.," Mr. Dillon said in a public speech. Soon afterwards he declared, in reference to America, that "full co-operation in war is none too much to offer to the friends who stood by us for two centuries and without whose friendship this nation would not exist today."

The second article appeared to be less popular, however, and can be found only in the *Toronto Star* and the *Ottawa Citizen*, the version reproduced here. Installments 3–5 never appeared. Whether they were never written, were banned by the censors, or CP just wasn't interested is unknown.

ning to alarm not only his girlfriend and his gaolers, but his fellow internees as well. What could they do to help? Was is it their fault? Or was it his internment? Was Jack cracking up, losing his mind? Was he simply the first, with others to follow? How much worse would it get? Clearly this had nothing to do with the articles. So what if some slick wise-ass reporter from some fancy New York magazine beat him to the punch? It wasn't the end of the world.

Everyone was concerned, and none more than Colonel McNally. Though Jack was not about to see him, no matter how often the colonel showed up at his hut. No matter how often that kindly concerned face appeared before him, with those round horn-rimmed glasses, bleeding compassion and understanding with every breath. No matter how often the colonel asked. That wasn't part of the plan.

The book! It had to be the book! They knew it was a stupid idea, an unattainable goal. Only a fool would try it. History never breathes, or lives. It dies. All history. Only the present matters, and now Jack had a full one hundred pages of hand-written foolscap, single-spaced, both sides, first under his bed, then in the top drawer of his dresser, and finally strewn about the floor. One hundred pages, can you believe it! And it was useless, totally useless! Who would ever read it?

A text on Irish history for chrissake? Like no one had ever done that before! Hello!

"Quick, Sir! Quick. Mr Calder, sir! There is something the matter with Mr Calder!" Brady and Girdlestone looked up from their cards, hardly believing. Could he have finally done it? Had they waited too long?

Quickly they jumped up from the table, and rushed over to Jack's room. In a drunken frenzy, Jack had picked up his typewriter, and was now holding it high above his head. Everyone stopped, aghast.

"I can't stand it anymore, he shouted. "I JUST CAN'T STAND IT."

Suddenly, with a mighty surge of strength, Jack threw the typewriter violently against the wall. The antique smashed into pieces, keys scattering everywhere. He then picked up a pile of foolscap from the floor, and began tearing it furiously to shreds.

"Your manuscript, Jack," shouted Girdlestone, "Don't do it! Quick. Get the colonel, get the colonel."

Within a moment two guards from the police hut had appeared in Jack's room. By now he'd overturned his dresser and was ripping things from his walls. Girdlestone was the first to intervene.

"What's the matter, Jack?" he asked. "What's the matter?"

"Why did she leave me? Why? Couldn't she have waited?"

The two Irish guards, terror in their eyes, looked down on Jack's bed. There lay the letter from Ann. They knew about it, of course, so diligent was Commandant Mackay and his lackeys. But they didn't think it would be this bad.

"I CAN'T STAND IT ANY MORE," shouted the former reporter again, before collapsing on the floor of his hut – a bottle of empty iodine rolling across the floor.

"By Jesus," said the guard. "He's killed himself! Quick, get the colonel, get the colonel! Calder's killed himself! Calder's killed himself! He drank iodine!"

It was nearly midnight, and the other officers and NCOs began appearing, beside themselves with grief. Every one knew Jack would do it eventually: it was only a matter of time. God knows the signs had been there. They just hadn't known when.

She had had enough of that silly book, she wrote. She loved him like no other. She wanted to live with him for the rest of her life. She could wait, she would have to. But she had had enough of that silly book. What a waste! What a waste!

"Quick, lads," said Corporal Smeaton, another Irish guard who had just arrived. "Get an ambulance and help me get Mr Calder to the hospital. He looks bad."

It was the least they could do.

Within seconds Girdlestone had jumped forward to lend a helping hand. The four then picked the stricken reporter up off the floor, now semi-conscious, and carried him out. Yellow stains appeared on Jack's mouth, and he began to retch violently.

"Quick, quick," shouted Girdlestone, as they arrived at the police hut. "He might die if we wait for the ambulance. Take the truck instead." Girdlestone was pointing desperately at an army lorry parked beside the main gate. "He swallowed iodine! He tried to kill himself!"

The guard at the police hut watched as they pushed their way through the gate. "Wait, wait," he said, waving a blank parole form in the air.

"Honestly, O'Reilly," replied Girdlestone dismissively. "How can you bother us at a time like this?"

"But sir," the junior guard replied to an obviously distraught Corporal Smeaton, "Sub-lieutenant Girdlestone hasn't signed his slip."

"Very well, very well," replied the corporal, as Girdlestone quickly filled in the form. They then carried Jack the rest of the way, past the colonel's hut and out through the main gate. Soon the former reporter was flat on the bed of the truck, writhing violently in pain.

"If we'd only known," said a new guard who had joined them from the duty hut, making a total of six. The guard worked with Commandant Mackay and knew about the letter. "Couldn't we have broken the news to him gently?"

"Quick, quick," shouted Girdlestone to the stunned driver. "He's going to die." Jack was now turning pale, obviously weak from the poison. The truck lurched forward as the six guards and Girdlestone sought to comfort him. "God, he's going to be sick again," said the New Zealander, as the others recoiled in horror. Quickly the driver pulled over. "Here Jack," offered Girdlestone. "Let me give you a hand."

"Just a minute please, sub-lieutenant," said a suddenly cautious Smeaton. Were they being had, or not? But then fearing that the man might well die from whatever it was that he had taken for whatever reason – the corporal stepped aside as Girdlestone helped his friend out the back.

Jack looked about. It was cloudy and dark, and the boughs of the birch and maple trees hung low over the road. Just as the last guard released his grip, Jack bolted down the road.

"Stop him," shouted the guards, as three joined in hot pursuit.

Girdlestone, who had given his parole, could only watch.

Unfortunately, within a minute of two, the four guards had caught up to the stricken, resourceful reporter, and were now leading him back to the lorry. "What was that all about?" asked one of them sceptically, as the four tossed Jack into the back of the truck.

"Don't ask," replied Girdlestone. "He's obviously delirious." And who wouldn't be after losing the love of your life? "If you don't get him to a doctor, the colonel will have something to say about it, believe me."

The guards recoiled. They knew how much the colonel cared for

DE VALERA LEADING IN EIRE BALLOTING

Counting of Votes Is Expected to Continue for Two More Days

Dublin, June 23—(CP)—Incomplete reports from many of Eire's 137 constituencies today showed Prime Minister De Valera's Fianna Fail government polling heavily in yesterday's general election as compared with the opposition. Voting was under the proportional representation system and vote-counting was slow.

Up to early this evening the Fianna Fail had won three seats and William T. Cosgrave's Fine Gael one. The Labor party led by William Norton had yet to score.

The public is awaiting the returns from Eire's first election since 1938 with the same calm as it maintained throughout the short campaign and the balloting. Ballot-counting is expected to continue through at least two more days.

Early returns from Cork city showed De Valera leading, while Cosgrave led in County Clare.

In the last House the government had 76 members and the Fine Gael 41. Labor held 10 and independents seven. There were three vacancies at dissolution.

CONCEDE DE VALERA TO RETAIN CONTROL

Eire Government 'In', Returns Indicate—Mulcahy Defeated by Larkin

Dublin, June 24—(BUP) — The government of Prime Minister Eamon de Valera will remain in power, returns from Tuesday's election indicated today.

Three key members of the opposition, Fine Gael (United Ireland) party were beaten, among them Gen. Richard Mulcahy a party vice-president.

With over half the vote counted, the latest standings were: Fianna Fail, 45; Fine Gael, 16; Labor, 11; Farmers, 5, and Independents, 3.

The pre-election division was Fianna Fail, 73, Fine Gael 40; Labor 10, Independents 8 and six seats vacant.

The Fine Gael's proposal for a coalition government appeared to have been rejected by the voters and its loss of front benchers impaired its leadership, regardless of how many seats it finally wins.

The counting confirmed the re-election of John Ryan, minister of agriculture, last member of De Valera's government to be in danger. De Valera himself was re-elected without serious trouble.

Mulcahy, Ireland's first defence minister, unexpectedly was defeated by James Larkin, Sr., Dublin laborite who was banished after the long transport strike of 1912. Larkin and his son, James, Junior, were elected after a campaign from a prison they had leased as headquarters. The senior Larkin was once confined in the prison.

James Dillon, formerly vice-chairman of the Fine Gael party, went back into his seat by a narrow margin in Monaghan county. He had been ousted from his party office after a speech advocating the granting of Eire bases to the United States.

Toronto Star, 23/24 June 1943. Although a de Valera majority looked likely at that point, the Irish leader won only a minority government in the end. (Courtesy, CP, BUP) the man. Still, it did seem a little suspicious, letters strewn about on the floor of his hut, Calder's sudden recovery from the poison, all on the day of the national election. What was the guy up to anyway?

"I bet it has to do with de Valera's victory," the guard added. Maybe it wasn't a broken heart at all.

"He had a spasm, you idiot," replied Girdlestone. "Can't you see that?"

Everyone knew de Valera would win, after all, even the internees. It was hardly a reason to go and kill yourself. No, that wasn't the sort of sympathy Jack was after.

Soon they were at the camp hospital. After Girdlestone explained to the resident doctor that Jack had taken some iodine – had actually tried to kill himself – everyone got concerned again and scrambled around trying to find an antidote. Jack himself suggested that sodium thio-sulphate might do the trick, having taken some earlier in the day, though by then he had thrown most of the iodine up and didn't really need it. To be on the safe side, they nonetheless transferred him the following day to St Patrick Dun's Hospital in Dublin, where he was seen by a couple of doctors, Dr F. Gill, and a specialist, Dr H.J. Eustace, each with sympathy both for him and his cause. This was at the suggestion of John Kearney, the Canadian high commissioner, who was equally sympathetic, having been supportive of not only Jack but of Keefer and Fleming as well. In fact, Kearney had known of the Canadian Press articles for weeks. He had also met Ann Mitchell several times. He knew not only of the letter but that the whole thing had been set up by Jack, with Girdlestone's help, months ago.

But it wasn't Mr Kearney's sympathy that Jack was concerned with. No, it was the colonel's. For when it came time to make a decision over Jack's fate, it would likely come down to what the colonel thought. And to no one's surprise, at least no one who knew the man, Colonel McNally recommended releasing Jack Calder the following day, with a couple of supporting letters on file. Jack had wanted to appeal to what he viewed as that most distinguishing of Irish traits, deep sentiment and excitability. And, quite to his surprise, it had worked. The colonel had bought it.

As if the colonel was doing Jack a favour.

DR. H.J. EUSTACE.

30, FITZWILLIAM SQUARE,

DUBLIN.

COPY/

29th June, 1943.

The High Commissioner for Canada,
92 Merrion Square,
Dublin.

Dear Sir,

At the request of Mr. F. Gill, F.R.C.S. I saw to-day Mr. Calder of the Canadian Air Force at Sir Patrick Dun's Hospital.

This man attempted to commit suicide by drinking Tincture of Iodine the previous day. He has now recovered from the immediate effects of his attempt and his life is not in danger.

He is very depressed, freely expressing ideas of hopelessness and lack of interest in his future. He still has markedly suicidal tendencies.

I understand that he has been interned for about 20 months; during part of this time he has had parole. He is now so depressed that he has little interest in the extra liberty and freedom that parole confers, and at the same time feels unable to continue the literary work he was doing when in camp.

In my opinion this man's depression is purely the result of his present position. His condition is serious; he requires constant supervision as he will certainly repeat his suicidal attempt. If he returns to camp life, the causes which precipitated his attack will be continued. If he is treated in a mental hospital he would require the strictest supervision, which would hinder his chances of recovery, though possibly preclude attempts to escape; probably too strict a regime would increase his depressive symptoms and lead to a chronic psychosis.

While he is urgently in need of skilled psychiatric treatment, to attempt in in his present position as internee would be very difficult, and the constant risk of further suicidal attempts would militate against any chance of success.

In my opinion the wisest course would be to transfer this man to a military mental hospital under the direction of his own countrymen, to receive the proper psychological treatment. If he is kept in this country, he will either commit suicide or develope a chronic mental illness.

Yours faithfully,

Sgd.) H.J. Eustace.

M.B. D.P.M.

Psychotherapist to the Adelaide Hospital, Dublin.

Consultant Psychiatrist to Sir Patrick Dun's Hospital
Dublin.

While the incident leading to Pilot Officer Jack Calder's release from the Curragh on 2 July 1943 generated a number of reports, only two are reproduced here: A medical opinion, if you could call it that, above, offered by Dr Eustace, and a memo from John Kearney, the Canadian high commissioner, opposite. Courtesy Irish Military Archives, Parkgate, Dublin; Department of External Affairs, Ottawa.

Airmail (confidential)
NO. 88

July 2, 1943

The Right Honourable,
The Secretary of State for External Affairs
Ottawa, Ontario, Canada

Sir,

1. I have the honour to refer to my telegrams no. 27 of June 29th, and no. 28 of June 30th concerning Pilot Officer John Philip Sargent Calder.

2. On the 29th of June I learnt that Pilot Officer "Jack" Calder had been transferred from the Curragh Internment Camp to Sir Patrick Dun's Hospital, Dublin, on the 28th of June. He was suffering from the effects of having taken iodine. It appears that, fortunately, after having taken the iodine he became nauseated almost at once, and quickly evacuated the poison. Calder had apparently been depressed for some time, and for weeks he has not taken advantage of parole but remained in camp working very hard on a book which he is writing. I am told that during this period he took no exercise, nor relaxation.

3. I visited Calder at the hospital and found him resting fairly comfortably. I next saw Mr Joseph P. Washe, Secretary of the Department of External Affairs here, and informed him of the matter. I asked him to have Calder medically examined with a view to procuring his permanent release from internment. Meanwhile I had seen Dr Gill who was attending Calder and he told me he would recommend that he (Calder) be released and repatriated. Mr Walsh stated that he would accept Dr Gill's statement in lieu of further medical examination, and suggested that, if possible, I procure an additional medical certificate. He informed that insofar as his department is concerned he would, on the strength of such certificates, recommend Calder's release but that he also had to satisfy the military authorities. The certificates in question were forthcoming, and both Dr Gill and Dr Eustace, an alienist, were of the opinion that Calder's was a mental case – that he had definite suicidal tendencies and that a return to internment might be fatal. As I informed you by telegram, Calder's release from internment was granted today.

4. Following Calder's release I arranged with Wing Commander Begg, who is Air Attache to the United Kingdom Office in Dublin and who is in command of all allied internees in Ireland, to take care of the case, and he is providing two male nurses who are coming to Dublin from Belfast to escort Calder out of the country.

5. I omitted to mention that on learning of Calder's case, I got in touch with Sir John Maffey, the United Kingdom Representative in Dublin, who said he would appreciate it very much if I would take charge of the case in so far as release from internment proceedings was concerned.

6. The United Kingdom Office in Dublin is very anxious that no publicity be given to the Calder incident, because they are endeavouring to procure the release of a Polish internee on medical grounds furnished by Doctors Gill and Eustace, and it might prejudice the chances of such release. Calder had made many unsuccessful attempts to escape from the internment camp, and whether or not this last episode is another daring, dangerous and disguised attempt is something which only Calder can conclusively answer. Calder, as you may know, in civil life, is in the newspaper business, and should he return to Canada it would be undesirable to permit him to write newspaper articles, or otherwise appear in the limelight, especially quickly following his release, on the ground of mental impairment. For his own sake I think it would be desirable that no publicity be given as to what was diagnosed as a mental condition, and that, in so far as it might be necessary to explain his release from internment, it should, I suggest, be simply said that he was released because of ill health.

7. I thought it desirable to keep you informed by telegram with regard to Calder's case so that you would have information on file should his parents get in contact with your Department.

I have the honour to be,
Sir,
Your most obedient Servant,

John D. Kearney
High Commissioner for Canada in Ireland

The Polish internee referred to in paragraph six is Zimek, who was released a few weeks later. Whether it was necessary for Kearney to go on and suggest that it would be undesirable for Jack to write any more newspaper articles, a virtual death-blow to his career, is very much open to debate. Courtesy External Affairs, Ottawa

10

BUS A, PLEASE!

When Jack Calder was finally back in England, having as much trouble convincing the British doctors that he was sane as he'd had convincing the Irish ones he was crazy, Bobby Keefer was finally on his way back to the Big Show. He had managed to survive the winter in North Bay, waking up at dawn with the sprogs, to shovel the runway in his greatcoat and fur-lined eskimo hat and now he would be transferred to the Number 1 Photo Reconnaissance Unit in Benson, fifty miles outside London. Fleming, who had arranged the transfer, would be joining him there.

So, finally, Keefer would have his wish. He'd be flying a Mosquito, the Wooden Wonder, nicknamed for its sitka spruce, balsa, and plywood frame. At 400 MPH – over double the speed of a Wellington – no one could touch him in one of those.

The two met in Montreal in early July to plan the impending move. They would ferry a Liberator to Morocco, they decided, so that Keefer could get ticketed on a four-engine aircraft, which would come in handy following the war. Now a flight lieutenant, Keefer would then complete an eight-week refresher course in Scotland, at Dyce near Aberdeen.

It was as they were making these plans in a bar in Montreal in the second week of July that they picked up a newspaper. It was a copy of the Toronto *Star*. Half-expecting to see the Jack's name there anyway, given all the coverage he was getting with his various exposés of Irish neutrality, they now read that he had escaped. Or, at least, that he had been released for, as the newspapers made plain, he wasn't talking. And the censors weren't either. That was more good news, they decided. Whatever had happened, they'd find out soon enough. The important thing was that soon they'd all be together again.

CALDER MUM ON HIS ESCAPE

Got Out of Internment Camp in Eire but Does Not Say How

LONDON, July 9—(CP)—Flying Officer Jack Calder, who was interned in Eire for almost two years, is back in England today but it must not be said that he escaped. The former Canadian Press Editor won't talk and the censors won't permit the cabling of stories about escapes from internment.

Less than a month ago newspapers published a Canadian Press series written by Calder in his internment camp at the Curragh, 30 miles from Dublin. One of the articles explained that the internees are not permitted by Canadian authorities to escape when they are out on parole, which is frequently, but they may do so when they have returned to their camp.

In March, Calder told visitors "very confidentially" that he "planned something" for June. Now he won't say a word. Last year his friend, Pilot Officer Bob Keefer of Montreal, escaped from the Curragh camp but Calder was hooked on a barb-wire fence when a makeshift ladder collapsed.

Calder and Keefer bailed out of a Canadian bomber over Eire in October, 1941, and were interned.

Calder, who had served as Ontario editor and sport writer at The Canadian Press Toronto Bureau before he joined the Air Force in 1940. Made newspaper history in August, 1941, with a bylined story on a raid on the German pocket battleship Gneisenau at Brest. The story, first bylined spot story of the war by a Canadian flier, was used on the front pages of newspapers across Canada and in the United States.

CANADIAN FLIER GOT AWAY FROM INTERNMENT IN EIRE

Flight-Lieut. Jack Calder of Owen Sound Now "Safe in Great Britain"

FATHER IS NOTIFIED

Flight Lieut. Jack Calder, former Canadian Press staff writer, who has been interned in Eire since October, 1941, is back in England, according to word received by his father, Rev. A. C. Calder, Owen Sound.

A wire from the R.C.A.F. at Ottawa disclosed no details of how his release was effected, Mr. Calder said today. "It merely said, 'Pleased to inform you your son safe in Great Britain.'"

"Surprised? I could hardly believe my eyes," said Mr. Calder, whose two other sons are also serving in Canada's armed forces.

The case of his son Jack is considered the classic internment story of the war. It started when his bomber crash landed in a marsh in Eire. He and his crew were shaken up but uninjured.

As belligerents in neutral Eire, they were interned but granted privileges unknown to regular prisoners of war or alien civilian internees. They were eventually put on parole which permitted them to go outside the stockade without escort, go for long walks, to attend movies, and even to telegraph

FLIGHT LIEUT. JACK CALDER

articles, many of which appeared in The Star.

Twice he tried to escape. Two of his fellow-airmen internees Flight-Lieut. Grant Fleming and Bob Keefer, both Canadians, managed to get clear the first time. Calder didn't.

In many articles written since his internment, Calder has given Canadians an insight into conditions in Eire. Often his stories sparkled with Irish wit especially when telling how he visited the Eire parliament and various "pubs."

The two clippings appeared in the *Toronto Star* and Brantford *Expositor* on 8 July and 9 July 1943 respectively. Other versions of the wire story appeared in other dailies, so famous had Jack become. While it's not clear when, or why, the former reporter received his promotion to flight lieutenant, it presumably wasn't as a result of his release. Courtesy CP

The flight to Morocco was Keefer's first official mission in nearly two years. To say that he was rusty wouldn't be true for he had logged another 500 hours teaching up north, but all of those had been aboard smaller aircraft, Tiger Moths, Hudsons, or Sterlings. Operating a Liberator would be a new experience.

On 12 August 1943 he and Fleming left Montreal. They stayed overnight in Goose Bay, Labrador, and then set off for Prestwick, via Reykjavik in Iceland. After a brief rest in Scotland the two crossed the Irish Sea (successfully this time), skirting the coasts of France and Spain and landing in Ra el Ma, in French Morocco. There they stayed overnight again – and had a close call in the eastern part of the city at a market where Keefer, attempting to negotiate a shawl for Blink, one of the girls at Churt, encountered a crazed merchant with a long, sharp knife. Fortunately, he managed to escape with his life – and the shawl. They then moved on to Castel Benito, Cairo, Lydda, Habbaniya, and finally Karachi on 28 August. Returning to London aboard an American Air Transport Command aircraft on 3 September, they arrived well ahead of schedule. Although Keefer found the Liberator took some getting use to – especially the noise from the engines' new turbo-superchargers – he loved it. Flying with four engines rather than two seemed so sensible, as did the ghouli chits,[43] clutched firmly in his pants pockets.

Not surprisingly, the first thing that the two planned to do in England was to visit Jack. They wanted to pick up where they had left off, now that they'd all be back in the racket together. They both assumed that the Irish had kicked Jack out of the country for making a nuisance of himself – which had been the initial plan prior to Sam Boal from the *New Yorker* showing up, and before Jack had opted in the end to play on that unique combination of Irish characteristics, sentimentality and excitability, that Girdlestone would often speak of later. But neither was entirely sure how he'd managed it.

Returning to London, though, Keefer was sure of one thing: with all the pounding the city had taken in the last few years, it seemed hardly a shadow of its former self. While the Brevet Club on Berkeley Square was still there, for the first time Keefer saw whole families living in tube stations, sleeping on three-tier metal bunks,

44 Ghouli chits were offers of reward written in Arabic and payable by the king of England should the bearer ever get captured by an Arabic tribe, neutral or otherwise, ignorant of the 1929 Geneva Convention regarding the treatment of interned prisoners of war. Ghouli is RAF slang for testicles, based on the experience of some unlucky pilots who had their ghoulis cut off and sewn into their mouths during the last war, something to keep in mind, he'd say, if you ever go to Morocco.

Photo of a Liberator, similar to the one they flew.

Year 1943 Month	Date	Aircraft Type	Aircraft No.	Pilot, or 1st Pilot	2nd Pilot, Pupil or Passenger	Duty (Including Results and Remarks)	Single-Engine Day Dual (1)	Single-Engine Day Pilot (2)	Single-Engine Night Dual (3)	Single-Engine Night Pilot (4)	Multi-Engine Day Dual (5)	Multi-Engine Day 1st Pilot (6)	Multi-Engine Day 2nd Pilot (7)	Multi-Engine Night Dual (8)	Multi-Engine Night 1st Pilot (9)	Multi-Engine Night 2nd Pilot (10)	Passenger (11)	Instr/Cloud Flying [Incl. in cols. (1) to (10)] Dual (12)	Instr/Cloud Flying Pilot (13)
–	–	—	–	—	—	— Totals Brought Forward	41:05	67:05			44:05	355:45	68:10	7:10	141:00	31:00	38:05	21:40	116:05
Aug.	12	Liberator	BZ863.	F/L Fleming D.F.C.	Self	Montreal to Goose Bay, Labrador.							5:00						15
Aug.	(Aug-13) 12/13	–	–		F/O Dark Nav. F/Sgt. Smith R/O Sgt. Broughton F/E	(15) Goose Bay to Reykjavik. Iceland.							4:40			5:00			4:45
Aug	13	–	–	F/L Fleming Self	F/O Dark F/S Smith F/E Broughton	(16) Reykjavik to Prestwick, Scotland.						5:25							
Aug.	20/21	–	–	F/L Fleming Self	F/O Dark F/S Smith	(17) Prestwick to Raz el Ma, French Morrocco.						2:35			8:30				5:00
Aug.	21	–	–		Sgt Broughton 4 Passengers.	(18) Raz el Ma to Castel Benito, Libya.						6:20							
Aug	22	–	–	F/L Fleming Self	F/O Dark F/S Smith	Castel Benito to Cairo West Egypt.						6:35							
Aug	26	–	–		Sgt Broughton 4 Passengers	Cairo West to Lydda Palestine.						2:25							
Aug	27	–	–	F/L Fleming Self	F/O Dark F/S Smith	Lydda to Habbaniya Iraq						3:00							
Aug	28	–	–		Sgt. Broughton 5 Passengers.	Habbaniya to Karachi India.						10:00							
Aug	14	Dakota	U.S. Army Air Corps.		Self.	Prestwick to Hendon.											2:00		
					GRAND TOTAL [Cols. (1) to (10)] 814 Hrs. 50 Mins.	Totals Carried Forward	41:05	67:05			44:05	392:05	77:50	7:10	149:30	36:00	40:05	21:40	126:05

Certified that F/Lt R.S. Keefer has been checked as captain on Liberator aircraft.
Aug. 13, 1943 J.B. Fleming S/Ldr

Keefer's log recording the trip to Morocco and Karachi, and Fleming's certification.

draped with clothing, dunnage bags, and ripped canvas suitcases. He also couldn't help but be moved by the children and the old people pointing to the flag on his epaulette. "Hello Canada," they'd say, wearing tired, worn-out grins. Maybe he hadn't noticed it on his last visit, so happy had he been to leave Ireland. But he certainly did on this one.

At dinner that first night, with Uncle Wilks, Dode, and Blink, they got some bad news: Jack had been in a crash, and the poor sucker had nearly bought the farm. Dode and Blink had seen Jack once, after he had been released from the Halton hospital in London in

WRITER-FLIER CRASH VICTIM

Former Eire Internee Flt. Lt. Calder Injured

London, Aug. 15 (CP)—Flt. Lt. Jack Calder of Owen Sound, Ont., was dangerously injured on active service when the plane of which he was a crew member flew into the top of a mountain in Northern England, it is officially learned.

Calder, formerly on the editorial staff of The Canadian Press in Toronto, suffered a compound fracture of the jaw and was taken to hospital in a large city near the English-Scottish border.

Later reports indicated Calder was suffering from shock and exposure and that his leg was in a cast, but he had improved sufficiently to be transferred to another hospital. His condition was still serious.

The casualty list announcing Calder's injury disclosed for the first time that he now is a flight lieutenant. Calder went overseas in the spring of 1941 as a pilot officer and was interned in Eire for more than a year after parachuting from his plane over that country in October of the same year. Early last month it was learned he was in England, but there was no explanation of how he reached there.

Toronto Star, 15 August 1942, the only newspaper to report on Calder's crash. Obviously, his fifteen minutes of fame was up.

the third week of July before the crash, having finally convinced the British doctors that he was sane. He seemed fine.

A woman, Ann Mitchell, was with him, added Dode, at the hospital, right there by his bed. She remembered Ann as a nice-enough girl, and Jack, as more than a little smitten.

"Dode, this is Ann, the girl I was telling you about," he said, suddenly quiet and soft-spoken, as if that had ever been part of his personality. Jack then told her that he and Ann had decided to get married. What happened over there in Ireland? she suddenly wondered. She thought he was getting tortured every day. Bobby had never mentioned anything about another woman.

Keefer flinched. Of course he felt responsible, apologising profusely. It was because of the big-build up he'd given Jack, he figured, that his cousin must have gotten her hopes up. Dode had to be forgiven for thinking that the guy had been scarred for life, as Keefer joked, by their hardships on the Emerald Isle. And then three weeks later the crazy bastard goes out and nearly gets himself killed on a cross-country training mission in an Anson, a small two-engine bomber that none of them liked. The plane had hit the top of Green Gable Hill on an all-night training run and suddenly Jack was back in the hospital, only this time he really deserved to be there. He really would be scarred for life, like London, less than a shadow of *his* former self. It had happened up north, she said, and two of the five had died, including the pilot.

Jesus Christ, thought Keefer, he'd only been back three weeks! What is it about these escapees? Do they all buy the farm within the first year? Still, Jack would pull through, Dode said. He was banged up, that's all. Had his face rearranged a bit. He was still on the bus, she said.

A Mossie MK IV B

The following day Fleming and Keefer got their marching orders. Fleming would be going straight to Benson, while Keefer would train briefly with the 418 Squadron, at Ford in Sussex, and then up to Scotland. The next day, Keefer was out in a dual control Mossie F II, with another friend, thinking, "Christ, at least I'm up here." And what a delight! While the typical twin-engine British aircraft such as the Wellington or the Anson did 120-140 MPH at 20,000 feet, these beauties cruised at 390 at 40,000! In the next two years Keefer would fly eight different versions of the Mosquito, but he remembered that first flight as if it were yesterday. The 418 was busy acting as fighter cover for Bomber Command sorties on the Nazi buzz bomb factory near Peenemunde that August, a factory that had been identified by photos taken earlier in the summer by one of the photo reconnaissance versions of the Mossie, the PR IV. (There were several bombing versions as well.) Indeed, the Mosquito, was a highly versatile aircraft and without it the allies might not have won the war.

The next week Keefer packed up his belongings in a small gunny sack and flew to Aberdeen and the number 8 OTU (Officers Training Unit) at Dyce. He missed Jack, who had moved to a hospital in East Grinstead, Sussex, and looked forward to seeing him on his next trip down. In Dyce they had the same Mossie trainers as at Ford. He even ran into his former squadron commander, Freddy Ball, the Old Man at the 103 in Lincolnshire who was now Dyce's CFI (chief flying instructor). Much of Keefer's time there was spent on the ground, learning about photography theory, interpretation and short-focal-length cameras. He also picked up a few tips on how to outrun the fastest German fighters, and how many things could go wrong on a photo-recce mission as compared with bombers and fighters, where you just kicked the tires and climbed aboard. Still, the speed of the Mossie eliminated a lot of worry. Most of the time German fighters could see you, or at least your contrails, but Mossies were so fast, and could go so high, that they couldn't do anything about it. This was a good thing, since most of the PR Mossies were unarmed, which is what gave them the extra speed advantage. Given the two American-made Merlin engines, Keefer needed just to fly the blessed thing, rely on his navigator, and take photographs with one of five cameras that he worked with a joystick. And the best news was that as soon as the new two-stage engines came out, they'd be even faster.

Finally, at the end of the training course, Keefer rejoined Fleming at Benson. It was the number one photo-reconnaissance unit in the war, and Keefer was delighted. He was so busy learning new things that he missed Jack in the hospital on the way down to Benson too, but swore to stop by on his next leave. He and Fleming were formally attached to the 540 squadron. Over the next several months, the two flew everywhere, with everyone, in their spiffy Mossie IXs with pressurized cabins! This continued well into the following year, 1944. Photo-reconnaissance from England was done mostly from three bases: Benson (for southern Germany, Holland, Belgium, and northern France), Wick (for Norway and northern Germany) and Saint Eval in Cornwall, (for the rest of France and Spain). So he spent his time bouncing from place to place, visiting all those wonderful countries, soaring high above all that death and destruction below, taking pictures! What a pleasant way to spend the war, he later joked. He felt just like a tourist. He was so busy, learning so much, that he barely had time to think of his friend.

While Calder was recovering from his injuries, and Fleming and Keefer were off taking pictures, Girdlestone, Covington, Brady, and the rest of the crew were still back on the Curragh, looking for a way out. The tunnel was progressing, albeit at a very slow rate. Several of them were now working on the project and by the middle of September they had tunnelled ten or twelve metres, leaving only fifty to go. This meant that they should have revised their initial time estimate from D-Day to V-J Day – had they known about either.

Other problems associated with the project at this stage – aside from a manpower shortage – included the lack of a proper aeration system, and because of it, an increasing infestation of fleas. There

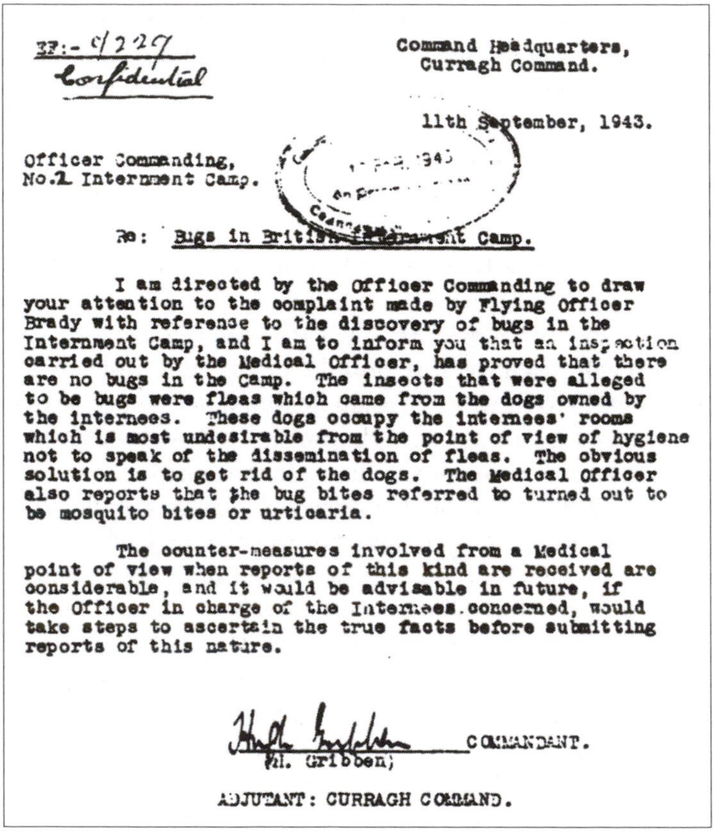

CF:- 9/229
Confidential

Command Headquarters,
Curragh Command.

11th September, 1943.

Officer Commanding,
No.2 Internment Camp.

Re: Bugs in British Internment Camp.

I am directed by the Officer Commanding to draw your attention to the complaint made by Flying Officer Brady with reference to the discovery of bugs in the Internment Camp, and I am to inform you that an inspection carried out by the Medical Officer, has proved that there are no bugs in the Camp. The insects that were alleged to be bugs were fleas which came from the dogs owned by the internees. These dogs occupy the internees' rooms which is most undesirable from the point of view of hygiene not to speak of the dissemination of fleas. The obvious solution is to get rid of the dogs. The Medical Officer also reports that the bug bites referred to turned out to be mosquito bites or urticaria.

The counter-measures involved from a Medical point of view when reports of this kind are received are considerable, and it would be advisable in future, if the Officer in charge of the Internees concerned, would take steps to ascertain the true facts before submitting reports of this nature.

COMMANDANT.
(H. Gribben)

ADJUTANT: CURRAGH COMMAND.

The Army's reply to concerns raised by one of the guards about the fleas. Excavations on the tunnel stopped briefly while the cause of the problem was determined. Courtesy Irish Military Archives

were so many dogs in the kennel, gifts from either the colonel or one of the local hunters, that a serious waste management problem had arisen, stinking up the place in the worst way. Fleas were everywhere, including underground and, with a tunnel circumference of only two feet, they didn't have a lot of room to itch down there. To make matters worse, they nearly lost Brady one afternoon, in the tunnel, after he had passed out from breathing too much of the thin, fetid air. Running a hose down solved the intake problem, but they still needed a way of pumping out the carbon dioxide. They solved that by buying a bellows as a wedding present for Sgt Tees, who had married that month, and then by promptly stealing it back.

But their ultimate undoing was Ward. Having resumed his post following Fleming's departure, Ward found out about the tunnel on a freak visit to camp to meet the diplomats. After Calder's release, there had been a second internee released on "compassionate" grounds – Jan Zimek, the Polish-American pilot who had suffered the fractured skull at the hands of the two Germans that spring – and the three diplomats, Maffey, Kearney, and Begg, had come to camp to "explain things." Such as why, if the Irish government was suddenly feeling so compassionate, not let the rest of them go? Didn't they deserve compassion and understanding too? (Even if in reality Zimek's release had more to do with what the Irish and the English governments thought Zimek might do to the next German he found out on parole than any compassion for his medical condition.) In addition, there had been rumours of a camp closure ever since Jack's release.

"Bang on, Sir John," exulted Ward, applauding enthusiastically from the front row, after Sir John Maffey had once more addressed the assembled internees one afternoon in the non-commissioned officers' mess. Only by now, fewer remained in awe at his impressive speaking style.

"Thank you, Flight Lieutenant Ward," Sir John replied, no less enthusiastic than his number one supporter. "I can't say for sure 'when,' chaps, but the Irish have made it pretty clear that it will happen soon."

Not even Covington bothered to ask.

"Yes. The talk is about dividing you lads into two groups. Bus A will be the lucky ones. They will be taken to Aldergrove and sent back to their units, and Bus B well ..." the British representative cleared his throat; everyone was definitely interested in this part, "they will be taken to a new camp, Gormanstown, twenty miles north of Dublin." As if it was some kind of honour.

"So who will be the lucky saps on Bus B?"

"Yes, good question! Not entirely sure about that, yet. We're negotiating with de V." Good thing, Keefer wasn't there. He'd hated that. "My hunch is that Bus B will be those of you who arrived on Ops, while the rest of you will be on Bus A."

"We all arrived on Ops, sir."

"Yes, Covie, that might be true. But we are attempting to narrow the definition of Ops." By then the British representative had lost them for all time.

A: BUS -

NO.	RANK	NAME
P.O.557	Captain	Baranowski, Kaz.
973926	Sergt.	Barrett, Wm.
R/69652	Pilot Officer	Brady, Charles
42591	Flying Officer	Covington, A.R.
R/58474	Sergt.	Fowler, Guy Duncan
	Sub./Lieut.	Girdlestone, B.W.
749495	Sergt.	Harkell, Robert Geo.
816145	Sergt.	Jefferson, George V.
483995	Sergt.	Karniewski, Stanis
	Flying Officer	Midgley, D.A.E.
911625	Sergt.	Masterson, James C.
581473	Sergt.	Ricketts, H.W.
	Flying Officer	Remy, Maurice
51946	Sergt.	Sutherland, D.
78644	Pilot Officer	Shaw, J.
551678	Sergt.	Todd, Norman
R66043	Sergt. Pilot	Tees, Roswell
1378792	Sergt.	Virtue, Alex
42472	Flying Officer	Welply, Denys
41501	Flying Officer	Ward, Leslie John
	Pilot Officer	Wolfe, Roland L.

B: BUS -

NOS	RANK	NAME
657301	Sergt.	Taylor, Donald Hugh
	Sergt.	Ross, David Eric Munroe
R64726	Flying Officer	Tisdall, Fred Wm.
1158678	Warrant Officer	Wakelin, James
968590	Warrant Officer	Newby, Herbert John
644074	Warrant Officer	Reid, David
137590	Warrant Officer	Diaper, Leslie
987421	Warrant Officer	Brown, M.B.
1266502	Flight Sergt.	Holloway, John Ed.
1365352	Sergt.	Slater, George

Courtesy Irish Military Archives

"Officers in bus A. NCOs in bus B. Is that it, sir?" Les Diaper, Keefer's second pilot, surprised just about everyone with that one. Though in the end Diaper was right. When they finally closed the camp, the following month, on 18 October 1943, posting the list the night before as if posting the final cuts for the football team, the only ones on Bus B were NCOs – however one chose to explain it.

Regardless, it was after the meeting about the two buses that Ward had discovered the Guinness crate. He then told the guards, the bastard.

Of course, you have to stay on the bus in the grander sense, you have to stay alive, for any of this to matter. Keefer could never explain where the expression "staying on the bus" came from, only that it meant what it meant, that it often came up during the war, especially in the air force, where fate, or luck, played such a role in

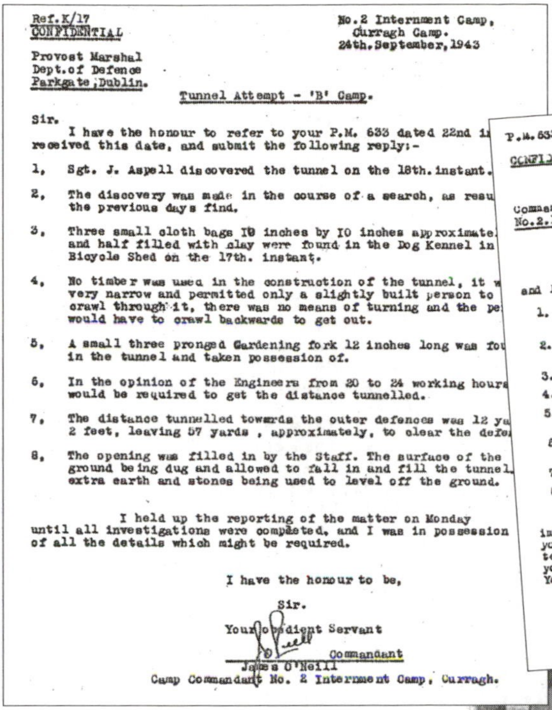

Ref.K/17
CONFIDENTIAL

No.2 Internment Camp,
Curragh Camp.
24th.September,1943

Provost Marshal
Dept.of Defence
Parkgate,Dublin.

Tunnel Attempt - 'B' Camp.

Sir.

I have the honour to refer to your P.M. 633 dated 22nd i[n]
received this date, and submit the following reply:-

1, Sgt. J. Aspell discovered the tunnel on the 18th.instant.

2, The discovery was made in the course of a search, as resu[...]
the previous days find.

3, Three small cloth bags 10 inches by 10 inches approximate[...]
and half filled with clay were found in the Dog Kennel in [...]
Bicycle Shed on the 17th. instant.

4, No timber was used in the construction of the tunnel, it w[...]
very narrow and permitted only a slightly built person to [...]
crawl through it, there was no means of turning and the pe[...]
would have to crawl backwards to get out.

5, A small three pronged Gardening fork 12 inches long was fou[...]
in the tunnel and taken possession of.

6, In the opinion of the Engineers from 20 to 24 working hours
would be required to get the distance tunnelled.

7, The distance tunnelled towards the outer defences was 12 ya[...]
2 feet, leaving 57 yards , approximately, to clear the defe[...]

8, The opening was filled in by the Staff. The surface of the
ground being dug and allowed to fall in and fill the tunnel[...]
extra earth and stones being used to level off the ground.

I held up the reporting of the matter on Monday
until all investigations were completed, and I was in possession
of all the details which might be required.

I have the honour to be,

Sir.

Your obedient Servant

Commandant

James O'Neill
Camp Commandant No. 2 Internment Camp, Curragh.

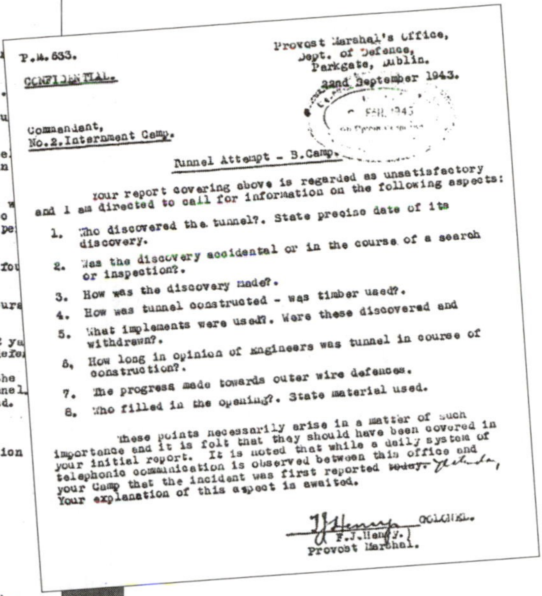

P.M.633.

CONFIDENTIAL.

Provost Marshal's Office,
Dept. of Defence,
Parkgate, Dublin.
22nd September 1943.

Commandant,
No.2.Internment Camp.

Tunnel Attempt - B.Camp.

Your report covering above is regarded as unsatisfactory
and I am directed to call for information on the following aspects:

1. Who discovered the tunnel?. State precise date of its
discovery.

2. Was the discovery accidental or in the course of a search
or inspection?.

3. How was the discovery made?.

4. How was tunnel constructed - was timber used?.

5. What implements were used?. Were these discovered and
withdrawn?.

6. How long in opinion of Engineers was tunnel in course of
construction?.

7. The progress made towards outer wire defences.

8. Who filled in the opening?. State material used.

These points necessarily arise in a matter of such
importance and it is felt that they should have been covered in
your initial report. It is noted that while a daily system of
telephonic communication is observed between this office and
your Camp that the incident was first reported today. yesterday,
Your explanation of this aspect is awaited.

F.J.Henry. COLONEL.
Provost Marshal.

Two reports, above and opposite, detail discovery of the tunnel later that day. Courtesy Irish Military Archives, Bruce Girdlestone.

Sgt Roscoe Tees's wedding was the only one to actually be held at the camp. Standing from left Karniewski, Holloway, Harknell, Sutherland, Jefferson, Ross, Fowler, Todd, Shaw, Ricketts, Reid, Thomas, Barnett, Slater, Wakelin, and Diaper. Sitting, from left: Masterson, Wolfe, Girdlestone, Tisdall, Tees, Eileen Lewis, Ward, Brady, Welpley, Midgley, and of course, Covington.

determining whether you lived or you died. To him, and to Jack and the others, it usually meant staying in God's good books, and the only way to do that was to write your mother every week, carry her replies with you wherever you went, treat your friends with kindness and respect, and pray. And even then there was no guarantee.

Still, he was as safe as anyone, Keefer realized, flying his Mosquito to the sun. An odd notion – flying to the sun. Nor could he say how many other flyers dreamt of that during the war. But he did, of hurtling forward beyond the earth's clutches at the speed of light, only to dissolve in a pyre of yellow and orange in one glorious flash of brilliance before reaching the heavens and saying good night. While the fact remains that it's impossible – that all aircraft are limited by speed and design to a certain height, whether 43,000 feet or Mars – it always seemed possible to him. Maybe that's why he left flying as a young man. There's nothing worse than a guy in a midlife crisis going out and buying a Cessna.

Anyway, the war continued, it never being a matter of choice for them, career or otherwise. Especially now, in 1944, in the fourth quarter, when the game was on the line, when survival was the only thing that mattered. The RAF stepped up its merciless pounding of German cities that Winter and Spring in what was euphemistically referred to as "strategic area bombing" which would eventually mean victory, he'd explain, as in jamming it down their throats big time. On 20 January 1944 hundreds of allied planes struck Berlin, dropping over 2,300 tons of bombs on military, industrial, and civilian targets. That attack, and a few more the next month, caused a furore even in London, where the House of Lords debated a motion of censure. While all were concerned with the ferocity of the attacks, and the abandonment of more humane principles of warfare, most understood in the end that nothing else would have worked. Later that year German V1 and V2 rockets, or buzz bombs, started dropping on London and would eventually claim 3,000 lives, with three times that number injured or maimed. Buzz bombs were liquid-fueled missiles that made a terrifying sound just before they landed, and he often felt safer in the air, at 40,000 feet over Germany photographing the factories that made them.

Still, it was a good year from his perspective, one that would eventually spell the end of the conflict. A well-earned respite for the lucky, and a much longer one for the rest.

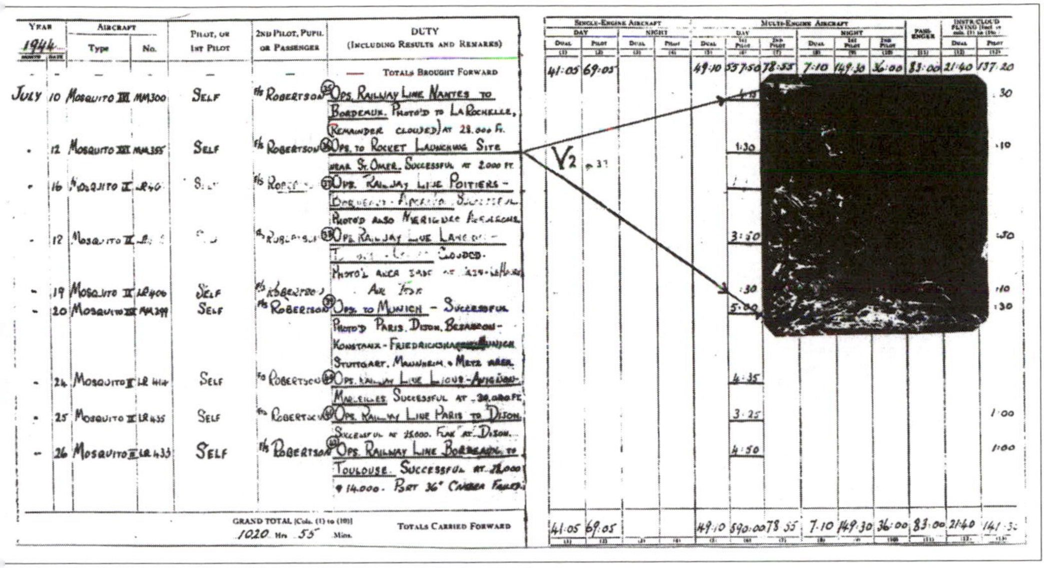

Year 1944 Month	Date	Aircraft Type	No.	Pilot, or 1st Pilot	2nd Pilot, Pupil or Passenger	Duty (Including Results and Remarks)	Multi-Engine Day 1st Pilot	Instr. Cloud Flying Pilot
–	–	–	–	–	–	Totals Brought Forward	537:50	137:20
July	10	Mosquito XVI	MM300	Self	F/S Robertson	(35) Ops. Railway Line Nantes to Bordeaux. Photo'd to La Rochelle, (Remainder clouded) at 28.000 ft.	4:[illegible]	.30
"	12	Mosquito XVI	MM355	Self	F/S Robertson	(36) Ops. to Rocket Launching Site near St. Omer. Successful at 2.000 ft.	1:30	.10
"	16	Mosquito IX	[illegible]	[illegible]	F/S [illegible]	(37) Ops. Railway Line Poitiers – [illegible] Successful. Photo'd also [illegible]	[illegible]	
"	17	Mosquito IX	[illegible]	[illegible]	F/S [illegible]	(38) Ops. Railway Line [illegible] – [illegible] Clouded. Photo'd area [illegible] Air Test	3:50	.50
"	19	Mosquito IX	LR406	Self	F/S Robertson		:30	.10
"	20	Mosquito XVI	MM399	Self	F/S Robertson	(39) Ops. to Munich – Successful Photo'd Paris. Dijon. Besançon – Konstanz – Friedrichshafen Munich Stuttgart. Mannheim. + Metz area.	5:00	.30
"	24	Mosquito IX	LR 414	Self	F/S Robertson	(40) Ops. Railway Line Lyons – Avignon – Marseilles. Successful at 30.000 ft.	4:35	
"	25	Mosquito IX	LR435	Self	F/S Robertson	(41) Ops. Railway Line Paris to Dijon. Successful at 25.000. Flak at Dijon.	3:25	1:00
"	26	Mosquito IX	LR433	Self	F/S Robertson	(42) Ops. Railway Line Bordeaux to Toulouse. Successful at 26.000 + 14.000. Port 36" Camera Failed.	4:50	1:00

V2 p. 33

Totals Brought Forward: 41:05 | 69:05 | 49:10 | 537:50 | 78:55 | 7:10 | 149:30 | 36:00 | 83:00 | 21:40 | 137:20

Grand Total [Cols. (1) to (10)] 1020 Hrs. 55 Mins. Totals Carried Forward: 41:05 | 69:05 | 49:10 | 590:00 | 78:55 | 7:10 | 149:30 | 36:00 | 83:00 | 21:40 | 141:3[illegible]

Keefer's log, July 1944, recording his many missions photographing the buzz bomb factories along the railway lines near Toulouse.

Just prior to Jack's release from the Queen Victoria hospital at East Grinstead in Sussex, Keefer finally got around to visiting his friend. It was in mid-February 1944, on Valentine's Day. He was always ashamed to admit that it took him so long, not only because of his fear of hospitals but because of what Girdlestone had intimated over Christmas. He and Fleming had met the New Zealander in London at the Brevet Club on Berkeley Square one night. While everyone had said Jack's injuries were a broken jaw and a compound fracture of the leg, the truth was worse than that. Otherwise Jack would never have found himself at the Queen Victoria hospital in East Grinstead.

East Grinstead was famous during the war for treating the worst RAF cases imaginable, not just scars or minor fixups, but major reconstructions, with missing ears, noses, and even eyes – "tiny bleak, glistening marbles, with a look which was not one to write about," as one journalist put it. Keefer found it difficult to even look at such cases on the street, much less visit them in the hospital – only to discover that his best friend was one of them.

Nevertheless it was his duty. And you don't turn your back on your friends.

He caught an early train from London, getting off at the East Grinstead stop, six or seven miles to the east of Gatwick airport.

East Grinstead, like many towns near London, was a small market town with half-timber buildings covered in ivy and a decidedly Tudor look. He found the hospital, roughly a mile north of High Street quickly, given all the friendly assistance he encountered along the way. He immediately understood why they called it the Town with a Heart, watching the ladies and younger women on the street, beside the bakeshop, or in the greengrocer casually smiling and chatting with his less fortunate colleagues, inviting them in for tea and cake. The flyers were all in uniform, disfigured in some cases beyond recognition. The previous summer East Grinstead had been struck by a 1,000 pound German bomb when its theatre was showing a children's matinee. The RAF patients were all that the women had left.

He'd try and smile too.

The next step was to call on his friend. Ascending the main stairway, he found Jack returning to his room from a commonroom at the end of the hall. He wouldn't look, he decided.

But there he was. A flicker of a smile creased Jack's face, or what was left of it, as they shook hands and slapped each other on the back. A deep cleft ran down one side, cutting through his eye socket and down his cheek.

"Don't worry, Bobby. One more graft and I'll still be better looking than you." Keefer didn't doubt it for a second. "I've missed you."

"Yeah, sorry about that."

Tears welled in Keefer's eyes. Still, it didn't take them long to relax again, given Keefer's pack of Newcastle Brown safely nestled under his arm. An hour or so everything else was more or less back to normal, cans empty, with Jackie FitzCalder and Babby O'Keefer living the life of Riley again. They kept at it all afternoon, covering so many things that he could scarcely recall them all later. All his questions were answered though, both as to the crash and as to what had occurred before that. They even talked about how great it would be if the two of them survived. What a reunion that would be! Coming over on the same boat, and then leaving on the same bus.

"Why didn't you just go back to Canada? Go sell war bonds?" Keefer had been waiting to ask that, thinking of all those pretty factory girls back home. They wouldn't leave Jack alone if they saw him like that.

"No way, buddy. I've got a few missions left in me yet. I'm going to join you guys. You wait and see."

"Why Jack? What's the point? The war's almost over, and you've got a career."

Christ the guy was stubborn. He'd written the warrant officer at Dumphries for a transfer to Coastal Command as soon as he heard what Keefer and Fleming were up to. Jack's problem at that point wasn't so much inexperience as the fact that a navigator's job had changed so much. The transition from a twin-engine mid-range bomber to a high-flying Mossie with a two man crew was daunting; far more so than for a pilot. First, he'd have to be ticketed as a wireless operator since the second man in a Mossie had to operate the radio in addition to his many other jobs. Then he'd have to learn the theory and practice of aerial photography, how to operate the cameras, etc. Then, at least according to this particular warrant officer, he'd need another 100 hours in the air. Once he'd gotten out, Jack started going up every chance he got, day and night, with Willie Panasik, whom they both knew from London. Willie was a top-notch pilot and no one thought he'd ever buy the farm on a training mission, taking one of his crewmembers with him, and putting Jack here. And that, Jack said, was why they'd made him a flight lieutenant. Though it should have happened long ago.

"And you, Bobby, congratulations, old boy!"

While Keefer was undoubtedly proud of his DFC, it wasn't as if Jack didn't deserve one as well. Hell, they all deserved them. Though Fleming, they acknowledged, fell into a different category.

"You know, Grant's the bravest guy I've ever known, Bobby. I still can't believe what he accomplished at the Curragh. With Shaw and Wolfe and Covie and all that, organizing everyone. The way he came into the camp and turned it around. We'd still be there if it wasn't for him. He had a hell of a way about him. And now he's out on Ops!" Unlike most other squadron leaders, Fleming was actually flying every day.

Jack was undoubtedly right. And yet, when Keefer thought back to that day, with Jack and him walking the grounds of the Queen Victoria hospital speaking in awe of their friend Grant Fleming, he couldn't stop thinking of how wrong he and Grant had been about Jack. Thinking that he was a reporter first, somehow, and a flyer only second. A gait from the screws left permanently in his leg, a face suddenly long and drawn and angular as he had never seen it before, and yet one eye, still sharp and bright and sparkling. He knew then that Jack was every bit as brave as he was, or Fleming. Or any of them. Even more so, since he, Jack, had had a choice. Though bravery can never really be measured in degrees, Keefer would say. It's either there or it isn't.

Indeed, Dr McIndoe told a Canadian Press reporter who had come at Jack's invitation that winter that he couldn't believe the courage and tenacity of the RAF men he had treated. Some had returned three or four times in the course of the war, dragged from the pyres of hell, to be rebuilt, reconditioned, for another operational tour.

"So what was that about, Jack? Why Considine and not you? Get scooped again?"

A sadness blurred the twinkle of his one remaining eye. "That was the deal, Bobby. That's why they sprung me. That I wouldn't write any more articles. That's why Mr Kearney did it." Keefer regretted the question; it was hardly necessary. Still, why had Kearney agreed to shut Jack up? Had that been necessary? "No Bobby, it's okay. That was the deal. I can't write any more, and I won't. That was my choice."

"Yeah, but that was only about Ireland, right?"

"No, buddy. It wasn't just Ireland. I'm not supposed to write anything at all. Not until after the war. That was the deal."

Jack's choice. But then he'd made a choice too, remembering Susan. They had all made choices.

They talked until well past supper. Jack and Ann were to be married in England, Keefer learned, as soon as the war ended. Though now he was surprised to hear that Jack wanted to stay there. That way Jack wouldn't have to worry about the censors remembering him, the Canadian ones. He'd become one of those nasty, nasty English reporters, reporting on Ireland, and the rest of Europe, from where it mattered most, from his armchair, with his one good eye, Ann at his side, and babies, lots of them, running around at his feet.

"Do it now, Jack. They'd release you. Go home!"

But no, Jack had too much fight left in him. He had spoken with a friend of theirs, he said, who had promised to get him into the Mossie light bomber unit at Oakington near Cambridge, the 571. And that's where they left it: talking about flying and the war, with Keefer trying to convince his friend to call it a day, to marry Ann

Flt. Lt. R. G. Keefer Winner of DFC

Ottawa, Jan. 16.—Flt. Lt. R. G. Keefer, 27, a Spitfire pilot attached to the RAF, has been awarded the DFC, according to official word received by his mother, Mrs. F. K. Keefer of Montreal and Ottawa. No citation was released at the time the award was announced.

Born in Ottawa, the son of Mrs. Keefer and the late F. K. Keefer, Flt. Lt. Keefer attended Upper Canada College while his parents lived in Toronto, and later attended Trinity College School, Port Hope. A graduate of the commerce and finance course at McGill, where he was well known as a rugby and hockey player, he enlisted in the RCAF after graduation, going overseas in April, 1941.

In December, 1941, he crashed in Eire and was interned. He escaped to England on Aug. 17 of that year, and was then sent home to Canada to instruct for a year. Since his return overseas he has been engaged in photo-reconnaissance work.

Ottawa Citizen, 16 January 1944. Once more they didn't get it quite right. He never flew Spitfires.

and settle down. Surely his friend had had enough of hospitals – and then Keefer feeling guilty for even suggesting such a thing. For they were all in it together to the bitter end, no matter what.

Which meant that Jack won in the end. He returned to Bomber Command and began missions with the 571 Pathfinders later that spring.

The next six months, from March to September 1944, saw Keefer, Fleming, and Calder racking up the missions and waiting for that momentous day when Europe would be reclaimed and they could all go home. And while it was a difficult period, with the romance of flight having long since given way to the drudgery of a job, with so many empty chairs in the mess each week that they scarcely knew whose turn would be next, it was also rewarding. For they each had a role to play, knowing that they had already won the air war, no matter how long it took the grunts and groundpounders to wrap things up. They could have gone home, having each completed several operational tours. And maybe they should have, given what would happen in September of that year. But they had made a promise. And they were not about to quit while they were ahead.

So the spring and summer meant more missions, and more fun. After Jack was released from hospital, he joined the No.8 group of Pathfinders. He couldn't get in at Benson, but he was now flying Mossies too, equipped with 4,000 pound bombs, and focusing primarily on night raids on German industrial targets. The Mossie

SUSSEX TOWN PATCHES ITS OWN BROKEN HEART THEN RESCUES FLIERS

Succors Burned Airmen Though Bleeding From Its Own Wounds—Nazi Bomb Snuffed Out Lives of Children—Unparalleled Job

By BOB CONSIDINE
East Grinstead, Sussex, England.

Toronto Star, 9 September 1943. The story that Jack would like to have written but couldn't.

bomber had a remarkable load capacity, carrying more bombs than a Boeing B-17 and providing further testimony to the strength of the original design. His squadron was also active in low altitude precision bombing, accomplishing numerous strikes on military targets, including the Gestapo headquarters in Denmark and the Philips Radio Valve works in Eindhoven. He took part in the celebrated bombing of the Amiens prison which allowed 250 prisoners from the French resistance movement to escape execution. With a light bomber that was faster than all but one of the German fighters, the RAF now had a critical advantage, which was an important factor in their eventual victory. Still, every mission was dangerous, flying deep into Germany marking targets for the swarms that followed.

As for Fleming, Keefer saw him more frequently, since they were both still flying out of Benson, though usually for briefer periods. (When he and Jack got together at the Brevet club, it was seldom brief.) They were both in PR versions of the Mossie, which were the fastest of the bunch, especially the XVI. Keefer had met a woman whom he liked, Mary Oldfield, who worked in the Ops room at Benson. Soon the six began hanging out together on leaves: Keefer and his new friend, Jack and Ann, and Fleming and his Irish girlfriend, who had settled in England for the balance of the war. They too had decided to tie the knot when it was over. They planned to settle in Alberta, on a ranch with real horses, a challenge for any daughter of the Irish Ascendancy. Mary Oldfield had a rich uncle, and one day the six of them climbed into his Armstong-Siddley convertible for dinner at White Hart in Nettlebed, where the Oldfields lived. They even planned a reunion one day, in Kildare.

As for Keefer, he flew so many missions that year that he could scarcely recall any of them. He suspected that he had been singled out for special treatment, or punishment, by the Old Man, Freddy Ball, because of their previous friendship at the 103. His frequent companion was Robby Roberston, his navigator, a flight sergeant from London and the one constant whenever he flew a two man Mossy, beginning when he came to Dyce and ending with VE Day in May of 1945. His most successful missions in that period were flown in the weeks immediately prior to D Day in June. First he covered an explosives plant at Angoulene, France, before and after a successful strike by Lancasters. Then he shot the entire coast, from the top of Denmark to the tip of the Spanish border, in preparation

A photograph that Keefer took of the Poudrerie Nationale works factory at Angouleme, sixty miles northeast of Bordeaux, an hour after it was struck by Lancasters. The factory made nitrocellulose used in the manufacture of powder.

for the "big landing." In the weeks that followed he flew every day, weather permitting, with Robbie popping up and down into the nose to line up their two thirty-six inch cameras, which they used for photographing targets, or the two cameras with fourteen inch lenses that were used for mapping. He shot everything from railway and marshalling yards to U-boats and factories, 2,000 negatives a pop. That was his job, as handed down by SHAEF, the Supreme Headquarters Allied Expeditionary Forces in charge of overall planning. It was fun – wrestling Robbie for the film and then racing it back to Mary in the operations room, hands behind his back, film in one and roses in the other. However she guessed, she always got both. He remembered one day in particular, 10 May 1944. They started out in the morning near Annecy, south of Geneva, and then moved on to five other towns to the north, including Nevers, Chartres, and Caen. He shot 5,500 negatives in a twenty-four hour period, climbing high above the channel and out over the Mediterranean at 40,000 feet, covering the Gironde ports along the French coast and watching the ME 109s disappear below, eating their vapourized dust!

Seeing all the German U-boats off the coast of neutral Spain, hiding. Wondering when it would all end.

EPILOGUE

Of course it did end. Jack would be the first, writing a letter to his mother three days before his death, proving himself wrong.

I guess my father should have known – Jack and him, fools that they were, thinking that they could all go on dodging the bullet forever, beating the odds, enjoying the things that had gone wrong and then right again. Taking their Scottish, French, or Irish leaves. He would have preferred to have sent us to bed thinking otherwise, I know that. But maybe it was better this way, once we were old enough to understand.

"Better that you hear the truth than you get the wrong idea," he'd say, "we never really lived the life of Riley, children. Jack was a remarkable man, a tremendous loss for all of those who knew him – his thirst for life, his intellect, his charm, and his courage. The Brevet Club's best-liked face. That's how we coped, that's all."

It was a Pathfinder mission to Hamburg on 20 July 1944 that finally did Flight Lieutenant Jack Calder in. His Mosquito, an ML 984, was hit by flak east of their target, his pilot, David L. Thompson, explained many years later. Jack, wounded in both legs, couldn't reach the K dinghy behind him. As the wooden wonder burst into flames, he hit the silk anyway, but drowned in the Elbe Estuary west of Hamburg, near Brunsbuttel, in the North Sea. His body was found washed ashore by a German fishing boat six weeks later, before being interred permanently at the British Military Cemetery in Kiel.

Keefer knew by then that something was wrong. He'd just

An excerpt from Jack Calder's letter to his mother, 12 July 1944, mailed three days before his death.

> Well, I'm back because I wanted to come back to it. I wasn't born to failure and disappointment. Quite a bit of me was broken up last time. But the realisation of what is within us all wasn't broken, thank God.
>
> "For God hath not given us the spirit of fear but of power and of love and of a sound mind."
>
> Mom, a lot of the boys leave letters behind to be sent to their people if anything should happen to them. I never have written that sort of letter to you and never will. I feel quite strongly that I am not going to be killed; and, because we shall be travelling much higher and faster than I ever have before, we should be "safer." But I don't want to preclude all possibilities and I know that you can be told these things now: That I am very happy. That if we should be attacked I am better informed and more alert than ever before about getting out of trouble. That if I should go missing then I would want you to be very quiet about it - particularly when the newspapers phone - because I probably would be walking back to you. And if I should fail to get clear, I would want you to think of me as walking towards you anyway, for that is what I would want to be doing.
>
> There is no death, you know.
>
> Love to all, Jack.

returned from the French Morocco on another photo reconnaissance mission. And the last time he'd done that, Jack had cracked up his Anson over Scotland.

Uncle Wilks had even looked for something in the papers. But there wasn't anything.

And then it was Fleming's turn. In the third week of September, still waiting to hear about Jack's fate, Keefer and another friend were having a drink at the Brevet Club in honour of the former reporter, waiting for their squadron leader to show up. As they announced last call, Keefer and his friend got up to join a couple of chorus girls at the next table over. No luck there, and still no Fleming.

God, please, not him too, thought Keefer. The guy who had got me out of Ireland, who'd been with me eating dried turnip for four days in the Aga Khan's stables, the guy who had rescued me from North Bay and kept me safe and sound out of Bomber Command.

On 15 September 1944 Squadron Leader Grant Fleming and two others from Benson went out to photograph the railroads and marshalling yards of Munich. It was one of the longer flights, Munich, and always a high-risk target. The Germans were reported to have new Messerschmitts there that were faster than the original Mossies but with limited range above 30,000 feet. Keefer had heard this, but he still hadn't seen any.

That night he rang Mary Oldfield at home and asked her if she had heard anything. She said the rumour was that a PRU Mossie had signalled an S.O.S. over Switzerland at 35,000, and that it had been chased for fifteen minutes by one of these jets. She was afraid that it might have been Grant's.

A month later the Swiss Red Cross reported that a Canadian-made RAF Mosquito had crashed high in the Swiss Alps. Fleming's body was never recovered.

Not surprisingly, Keefer could never really describe the rest of the war, not after that, not after having his two best friends die within weeks of each other. He must have felt as if someone had suddenly pulled the rug out from underneath him, kicked him where it hurt the most. It might also have occurred to him that he would be next. Remarkably he wasn't, something which he could only ascribe to good Irish luck – without a drop of Irish blood in him. It was never a matter of skill, despite what you hear of the Billy Bishops and Saburo Sakais of the world (Sakai was a Japanese pilot who downed sixty-four Allied aircraft during World War II in his Mitsubishi Zero fighter, and we all know who Bishop was). Without luck on their side, they would have died too.

The following week the Old Man assigned Keefer to a Rolls Royce engine-handling course in Derby, north of Birmingham. There were only two of them in the course, and Keefer suspected later that the Old Man had given him a break to collect his thoughts. Keefer saw the stage-2 Merlin engines for the first time, an improvement on the original Mosquito design that would likely have saved both Calder and Fleming. He was also asked to carry sealed documents to the Belgian front aboard a Lysander, and then asked to ferry a couple of shuttles to Italy – pretty light stuff. On 6 November 1944, Keefer made his first trip in one of the stage-2 Mossies. In the six weeks that followed he flew them many more times, detailing the Karlsruhe-Heilbronn-Crailsheim railway on three successive days in mid-December and four consecutive missions over Christmas, from the 24th to the 27th. Eventually, his confidence was restored – if not in God or justice, at least in his ability to fly.

In March 1945, the 540 squadron at Benson was moved to Coulommier, twenty-five miles east of Paris. The weather was better there, and they'd be closer to the action. Keefer packed up and asked Mary Oldfield to marry him. When she wouldn't give him an answer, he said goodbye – for good as it turned out. In France, their food improved noticeably – after four years of English and Irish cooking he noticed – as did their accommodation, an old manor on the outskirts of the city that had been a brothel during the occupation. As he put it, at least now the girls got paid. His prime duty was to monitor the movement of Panzer Divisions by rail, which he and Robbie Robertson, still his navigator, did – flying mission after mission detailing the railways, allowing interpreters to locate strings of flat cars carrying the German Tiger tanks that had proved to be so successful in Africa. Only now these same tanks were being pounded mercilessly because of their photographs! He recalled a three and one-half hour run over Kitzengen, Nuremberg, Regensburg, a brief stop for lunch in San Severo, Italy, and then a race back to England, all of which won him a commendation during the final push to V-E day. And, of course, he celebrated with the rest, when the war finally ended six weeks later, on 8 May 1945.

And yet, the irony of it all was that the greatest anger he felt for the longest time was not towards Germany or Japan for the loss of his two closest friends, and for everything else, but rather Eire, this place that he would describe so that we could judge for ourselves, with us rarely the wiser. You mean Ireland, dad? You were a POW in Ireland? One of only a few countries to avoid the conflict altogether, and the only English-speaking nation in the world to remain neutral. In short, my father agreed with Winston Churchill, who, in a victory address to the English people on 13 May 1945, five days after V-E day, accused the Dublin government of "frolicking" with the Germans. "With a restraint and poise to which I say history will find few parallels," he said, "His Majesty's Government never laid a violent hand upon them, though at times it would have been quite easy and natural to do so." Churchill was still angered by de Valera's visit to the German embassy in Dublin following Hitler's suicide in April. And yet, it is clear today that Churchill's comments were unfair.

A portion of the last page of Keefer's pilot log – eighty-eight missions in all!

Year 1945 Month	Date	Aircraft Type	Aircraft No.	Pilot, or 1st Pilot	2nd Pilot, Pupil or Passenger	Duty (Including Results and
–	–	–	–	–	–	Totals Brought
[Ma]r	19	Mosquito XVI	NS 640	Self	F/O Robertson	(79) Ops. Railway Line Nienburg & Hameln–H[...] Soest. Latter success[...] Also 2/4 Bridge at Ar[...]
Mar	21	Mosquito XVI	RF 970	Self	F/O Robertson	(80) Ops. Railway Line R[...] Nurnberg – Regensb[...] Successful at 25,000. [...] D/A 3 runs. Landed at Sa[...]
[Ma]r	21	Mosquito XVI	RF 970	Self	F/O Robertson	(81) Ops. Return from S[...] No photos.
[Ma]r	22	Mosquito XVI	MM 399	Self	F/O Robertson	Air Test.
[Ma]r	23	Mosquito XVI	RG 139	Self	F/O Robertson	(82) Ops. Railway Line Gottingen – Bebra at 26,000.
[Ma]r	25	Mosquito XVI	RF 980	Self	F/O Robertson	(83) Ops. Railway Line Hannover – Bremen at 26,000. Hannover burning furiously.
Mar	30	Mosquito XVI	NS 693	Self	F/O Robertson	Benson to Coulom[miers]

Summary for MARCH 1945
Unit #540 Sqdn Coulommiers
Date 1.4.45
Signature R Keefer F/L

GRAND TOTAL [Cols. (1) to (10)] 1235 Hrs. 30 Mins.

Totals Carrie[d]

MESSAGE OF CONGRATULATION.

The appended copy of a message of congratulation recieved from 106 Group, under cover of letter 106G/14/Air dated 22nd. March, 1945 is forwarded for your information and necessary action.

for Group Captain, Commanding,
R.A.F. STATION, BENSON.

"A telephone message has been received today from S.H.A.E.F. stating that Brigadier Foorde wishes to express the admiration of S.H.A.E.F. for the most excellent railway sortie flown yesterday by Flight Lieut. Keefer, sortie No. 106G/4965.
2. It is stated that the 6th Panzer Army moved East six weeks ago and the results of two sortie enabled two armoured divisions to be located by the Russians and also provided information on the possible movements of the rest of the Units.
3. This sortie is in accordance with the high standard already set by No. 540 Squadron, and it is requested, therefore, this expression of admiration from S.H.A.E.F. be passed ~~xx~~ not only to the aircrew concerned, but to all the Squadron."

Courtesy, RAF

For one thing, the Irish have always shown compassion, and there's no reason they should have made an exception for Germans. On New Year's Day 1944, 150 German sailors were rescued by an Irish freighter. While all the survivors were interned in the new camp, at Tintown, just down the road from where G camp had been (both sides of the original camp were closed in October 1943), the local people set up a "Help the German Sailors Fund." They collected food and household items and coordinated "friendship" visits. And while most, if not all, of the allied internees felt bitter towards Ireland – my father's surviving friends, Covington, Girdlestone, Brady, and Wolfe, included – the fact is that Ireland did favour them, at least to the extent that it could, while still remaining neutral. Documents released after the war establish that not only did top level meetings between British and Irish military authorities take place from 1942 onwards, (making Churchill's comments even more surprising) but information was also exchanged about German spies on Irish soil, including a number of the Curragh internees. And, as Mr Kearney had suggested to them, in early 1942 permission was given for allied flyers to use Irish air space. Information concerning submarine and aircraft activity around Irish ports and weather reports was also exchanged. The favouritism is also born out in a summary look at the numbers at the Curragh (and later at Gormanstown, the new allied camp). During the war there were 112 RAF landings or crashes in Irish territory, and 20-odd German ones. While only one out of every five airmen on the allied side was interned, (60 men in total, after all the escapees and compassionate releases are included), all of the Germans were forced to remain, 80-some flyers plus the 150 new arrivals from the German navy. In addition, no Germans were released in October 1943, unlike the lucky ones aboard Bus A. And while Gormanstown, the new allied camp for those on Bus B, was formally closed 15 June 1944, a week after D day, the Germans remained in their camp until 13 August 1945, until Japan's surrender. A large number of German airmen married local Irish girls that month, after the Irish government, in a final gesture of benevolence, allowed them to remain. Had Ireland been any more overt in its support, Germany would no doubt have attacked. And the war would have turned out differently – at least from the Irish perspective.

And what then of the others, those brave men, officers and sergeants alike, whom Keefer met in Ireland, who returned to active missions each and every one? In some cases, unfortunately, little is known. John Shaw and Denys Welply, for instance, two of those released aboard Bus A in October 1943, both died on operations in 1944 but he never knew when or where. Flight Lieutenant John Leslie Ward, whom no one seemed to like, went on to serve the RAF with distinction as the CO of a transit camp in Karachi. Keefer saw Maurice Remy briefly in Paris in early 1945, (they had coffee, since Maurice was a teetotaller) but he never heard from Stanley (PO Stanilsau Karniewski), the Baron (Flight Lieutenant Kazimeier Baronowski), or Midge (PO David Midgely). On the other hand, he did keep up with Bruce Girdlestone. After his release on Bus A, the New Zealander served in the Pacific in 1944 and 1945, where he met and married a member of Lord Louis Mountbatten's staff. He went on to become a successful architect in Christchurch, New Zealand, after the war, garnering many distinctions, including that of co-founder (along with my father) of the Remarkables, an organization of former internees of the Curragh that also included Chuck Brady, Bud Wolfe and Aubrey Richard Covington, its honorary president. Remarkable not so much for surviving their internment as for having survived everything after that. Brady went on to become a flight lieutenant and DFC winner, marrying a WAAF and retiring to Kelowna, British Columbia, while Bud Wolfe retired from the US Army Air Corps as a Lt Colonel, and moved to Orlando, Florida. All told, Wolfe flew 900 combat missions and 12,000 hours for the US Army Air Corps. As for Covington, he served out his RAF career with distinction, retiring as a Wing Commander. And then he started serving again, this time in earnest, as proprietor and manager of the Woolpack Inn in the Cotswolds, near Stroud, south of Gloucester.

As for Keefer's crew, Alex Virtue, his tail gunner, was released from the Curragh just before its closure in October 1943. My father never knew why until he received a letter in the mid 1980s from Maurice Brown, his nose gunner, who said that Virtue had died of cancer in Dublin in early 1944, an illness that had been diagnosed the previous year. Brownie and his Irish wife, Brigit Devlin, left Ireland after D-Day and the closure of Gormanstown. He served briefly on Ops before returning and settling in Dublin the following year – the only one among them to do so. And of course Dalton,

Wolfe and Covington, Florida, 1983

their wireless operator, died on Ops in the fall of 1943, prior to the closure of the camp. Les Diaper, who also married at the Curragh, and was on Bus B, went on to serve the RAF in Burma, flying Dakotas in support of Wingate's Chindits, and then in Batavia, flying women who had been held prisoner by the Japanese back to England. That was a harrowing experience, he said. My father saw both Diaper and Brown in England in 1986 and, after a close vote, managed to get them in as full time members of the Remarkables, despite their rank. Old ways die hard.

As for some of the others, Slapsy Maxie, of "I Bombed the *Gneisenau*" fame, retired a squadron leader, and Freddy Ball, Keefer's first and last Old Man, at least in the air force, went on to become Air Marshall Sir Alfred Ball, KCB, DSO, DFC. Coxie, his first tail gunner, who bailed out during the St Elmo's fire Keefer encountered two days before landing in Eire, survived his stint in a German

Tel: 0452 813429

WOOLPACK INN
ACCOMMODATION AND BAR SNACKS

Slad
Nr. Stroud
Gloucestershire

Proprietors:
Mr. & Mrs. R. Covington

Covington's pub. Try the steak and kidney pie, its delicious!

POW camp and settled in England. Dode was heart-broken that Jack never married her but went on to marry someone else, which was probably better for her. Jack O'Mallye emigrated to America in the 1950s and died a short while later, and Corporal Lillis, one of the two men who arrested my father was alive and well in county Clare when I met him in 1994. Colonel Thomas McNally retired from the Irish army in the early 1950s, and a number of former internees continued to correspond with him. He, too, was reportedly saddened by Jack's death, showing how many lives were touched by the young reporter. By all accounts Colonel McNally was a fair and compassionate man – they could have done much worse! After Jack's release, the colonel requested a transfer and received it. The pugnacious Lieutenant Kelly went on to become a judge of the Superior Court of Ireland, believe it or not, which went on to become the Republic of Ireland, officially, in 1949. I'm sure Keefer was glad that he never had to appear in Kelly's court.

As for the girl friends, Mary Kelly, according to what she told me in Kildare, remained heart-broken over Grant Fleming's death. She never married. The Lawlors remained in Osberstown House after the war. Jim died of cancer in the 1960s and Tom of old age in the late 1980s. My regards to his lovely wife, Vi, whose hospitality and stories were most appreciated. As for Susan Freeman, she moved to England after the war and her father sold Ardenaude some years later; Keefer never saw her or Sheila O'Sullivan again (though he certainly dreamed of Sheila's sister more than a few times). And Ann Mitchell, who was reportedly devastated by Jack's death, recovered, as Jack would no doubt have hoped, only to die tragically herself in a hunting accident in 1952.

And, finally, Flight Lieutenant Bobby Keefer retired from the Royal Canadian Air Force after the war, having technically only been seconded to the RAF. He continued flying commercially, perhaps to purge himself of the sadness of losing Calder and Fleming and a host of other friends. He ferried aircraft out of Montreal for a while, just as he had in Dorval in 1943 with Fleming, establishing a world record, if only briefly, when he flew a twin-engined Beechcraft from Montreal to Cairo in just under forty-six hours. He later established a family and a life for himself in the import-export business there. Montreal, as an international city, was a lot busier back then than it is today; and like Kildare, it was also a lot more Protestant.

As for the romantic component, Keefer married an American in the end, my mother, Julia Roberts, from Macon, Georgia, near Atlanta. She came up one Christmas in the late 1940s to visit her brother at McGill and never returned. They had five children. While my dad had hoped to marry a WAAF – like his friends, Calder, Fleming, and Brady, who actually married one – he did just fine in the end.

Which brings me full circle, really, if only because one of the last times I saw my Old Man, he was back in the slammer once more, which seems as good a place to leave it as any. That was a few years ago now, at the Rideau Veteran's Home in Ottawa, which Sam Boal

Keefer, with his five demanding children, circa 1960. The sixth is a cousin.

Montreal-Cairo Hop in 46 Hours

WHAT is thought to be a new aviation record has been made by R. G. Keefer, D.F.C., of Montreal, former McGill rugby star, who has ferried a twin-engined Beechcraft 18 from here to Cairo in 46 hours. The aircraft left the factory in Wichita, Kansas, last Friday, stayed here overnight, left for Cairo Saturday morning, and yesterday was in service there.

Stops were made en route at Gander in Newfoundland Santa Maria in the Azores, Algiers in North Africa and landed at El Maza at Cairo.

Montreal *Star*, 6 August 1946.

of the *New Yorker* might have called a run-down summer camp, just what he'd called the Curragh. They looked alike in some ways – the Curragh, and the Rideau Veteran's Home – at least judging from the pictures.

Still, my dad seemed happy that day, when I hung my coat at the door and started walking down the hall. He smiled a lot, even winked at the nurses – though no one could understand what he said. He was one of thirty residents there – you weren't supposed to call them patients – all male, all World War II veterans, all telling war stories that no one ever believed.

At first they wouldn't let him in. They said his unit was reserved for former POWs, and he wasn't a former POW. To hell with that, we said.

It's an odd disease, Alzheimer's. He wouldn't eat what we told him to eat, and he spat out his broccoli, but he could still get through a meal. It was like he was a kid again. He'd play with his knife and fork for twenty minutes, clang them around in his glass of milk like he was looking for something (I wonder if he remembered me hiding my broccoli there), put his mashed potatoes in his pocket, and then all of a sudden something would click and he would clean up his plate like there was no tomorrow. Then he would start talking about his fourteenth mission – at least that was my guess. I remember him taking me to the window, where he'd watch the buses go by. For maybe an hour.

"Bus! Bus!" he said over and over.

He's wondering who I am and I'm thinking, Bus A or Bus B?

A rundown summer camp all right, only this time someone forgot to pick up the kids.

Keefer often said that Calder and Fleming were the two bravest men he ever met. Well, my dad was the bravest man I ever met. You see, I knew he knew. I knew he knew what was happening to him, and what would come next. There were fleeting moments of recognition – call it a son's instincts – for you could see the fear in his eyes. Especially when he looked at me. Maybe he thinks I'm him, the spitting image of his youth. That's enough to scare anyone.

The Rideau Veteran's Home, 1997

Bus A, at the Rideau. It wasn't all that bad. They took him out every day on field trips, and you couldn't find nicer people than the kind staff at the Rideau Veteran's Home.

Back to the window we'd go, for my sake, if not his. Outside it was clear, crisp and windy, a typical October day, a football day. He loved sports my dad, and was damn good at them. Did I tell you he was a great football star?

Suddenly I wanted him to be young again. I wanted to see his courage and strength, to hear his words, reassuring with their affection yet sharp when they had to be, to share in his spirit, to take me home, or to take me away with him. To provide me with opportunities he never had, and I never deserved.

Have I done a good job, dad? Have I been fair?

Have I ragged on the English too much, or the Irish too little?

Very well, Old Man, you've earned it.